UNDERSTANDING THE UNIVERSE

by

Robert E. Smolinski

DORRANCE PUBLISHING CO., INC.
PITTSBURGH, PENNSYLVANIA 15222

The contents of this work including, but not limited to, the accuracy of events, people, and places depicted; opinions expressed; permission to use previously published materials included and conclusions drawn from author-presented information; and any advice given or actions advocated are solely the responsibility of the author, who assumes all liability for said work and indemnifies the publisher against any claims stemming from publication of the work.

All Rights Reserved
Copyright © 2011 by Robert E. Smolinski
No part of this book may be reproduced or transmitted in any form or by any means, electronic or mechanical, including photocopying, recording, or by any information storage and retrieval system without permission in writing from the publisher.

ISBN: 978-1-4349-0843-8
eISBN: 978-1-4349-5597-5

Printed in the United States of America

First Printing

For more information or to order additional books, please contact:
Dorrance Publishing Co., Inc.
701 Smithfield Street
Pittsburgh, Pennsylvania 15222
U.S.A.
1-800-788-7654
www.dorrancebookstore.com

ACKNOWLEDGMENTS

I wish to extend credit to the following publications for verification of the latest data on parameters associated with the stellar bodies in our universe. They include:
- *Encyclopedia Britannica*
- *Life Nature Library—The Universe*
- *Atlas of the Universe*—Patrick Moore
- *Universe*—D.K. Publications
- *Cosmos*—Carl Sagan, Random House
- *Astronomy* magazine
- *Scientific American* magazine
- *The Big Bang*—Joe Silk
- *On the Shoulders of Giants*—Steven Hawking

UNDERSTANDING THE UNIVERSE
TABLE OF CONTENTS

Foreword ix
Introduction xi

Chapters

Chapter 1 - Our Solar System...1
Chapter 2 - Earth's Place in Our Solar System 62
Chapter 3 - The History of our Astronomers, 65
Scientists, and Mathematicians 65
Chapter 4 - Understanding Earth through Time 71
Chapter 5 - Galaxies - Distances and Types.......................74
Chapter 6 - Hertzsprung-Russell Diagram 77
Chapter 7 – The Beginning of Our Universe 80
Chapter 8 - The Red-Shift 84
Chapter 9 - Reference 90
Chapter 10 - Understanding Galaxies...............................111
Chapter 11 - Future of Planet Earth..................................113
Chapter 12 - Possibility of Life out There 114
Chapter 13 - Problem with Space Travel 115

Glossary 117

PREFACE AND DESCRIPTION OF THE CHAPTERS

I will start our journey to the stars and beyond by first describing in detail our Solar System, with data and pictures of the sun, planets, major moons, the asteroid belt, Kieper Belt, and the Ort Cloud. Chapter One is fairly extensive; I did this deliberately, so the reader can have at his or her fingertips data of our Solar System, for reference and to lay the groundwork for a better understanding of the following chapters.

Chapter 1 - <u>Our Solar System</u>
Reference of our Solar System with data and pictures

Chapter 2 - <u>Earth's Place in our Solar System</u>
This chapter describes my reasons why I think the earth is unique in the universe. I'm sure the readers have heard the numbers argument many times. Let me suggest they consider what I have to say to excite their imaginations.

Chapter 3 - <u>History of Our Astronomers, Scientists, and Mathematicians</u>
Here I present the names, dates, and accomplishments of our most famous astronomers, scientists, and mathematicians.

Chapter 4 - <u>Understanding Earth and Our Planets through Time</u>
In this chapter I describe the stages of earth's development, along with our other planets, from their beginning to present.

Chapter 5 - <u>Galaxies - Distances and Types</u>
Here I present galaxies, their types, physical size, compositions, and distance.

Chapter 6 - <u>The Hertzsprung-Russell Diagrams</u>
In this chapter I discuss the who, what, when, where, and how of the Heartzsprung - Russell Diagram and how it came into being. It's one of the most useful tools ever developed for astronomers and because it's related, I include a description of how a spectroscope works.

Chapter 7 - <u>The Beginning of our Universe</u>
In this chapter I present my reasons and arguments why I disagree with the Big Bang Theory, the use of the Scientific Method, and why I favor the Steady State Theory.

Chapter 8 - <u>The Red-Shift</u>
This is the chapter that motivated me to write this book. I present my theory, arguments, data, formulas, and reference material relating to the Red Shift, along with diagrams, tables, figure, and

pertinent ideas that relate to it, possible applications of my theory to other experiments, and empirical tests in the field of cosmogony and cosmology.

Chapter 9 - <u>Understanding Galaxies</u>

In this chapter I discuss my perspective on yet undiscovered physical laws, such as the light barrier, galaxies "winking out," dark matter and dark energy, the Steady State Theory, and the Big Bang Theory.

Chapter 10 - <u>Future of Planet Earth</u>

In this chapter I describe my visions of the future of our earth and address questions like how long I think the earth will last.

Chapter 11 - <u>Possibility of Life out There</u>

In this chapter, I give my reasons why I don't believe in UFO's, ET's, or other intelligent life from elsewhere.

Chapter 12 - <u>Problems with Space Travel</u>

In this chapter I discuss the reasons why space travel for humans is really <u>not</u> possible, given our present technology. This final chapter raises many issues and presents much food for thought which provides concepts for future inventions and discoveries not yet developed—many ideas that I have been thinking about—and provides the groundwork for me to write another book. But let's see how this one does first.

FOREWORD

As long as I can remember, astronomy has been an important part of my life. When I was a young man back in the 1940's, I would spend my leisure time at the Adler Planetarium, in downtown Chicago. I spent many years studying the astrolabes, telescopes, and ancient instruments used by early astronomers and navigators dating back hundreds of years. Of particular interest was watching the star show projected on the inside dome of the building by the Ziess Planetarium and listening with great interest to the lecture's description of the heaven we were watching. Those many years I spent at the Adler Planetarium, I believe, were responsible for laying the groundwork and for my interest in the subject of astronomy, which was to last all my life. As time passed I was married, and my wife Elaine and I raised two daughters, Linda and Laura, here on the northwest side of Chicago, in the Portage Park area. The girls attended St. Bartholomew Catholic School, and somewhere around third or fourth grade asked if I would take them to get their first library cards. We did, and while we visited for the next few years, I became reintroduced to my favorite subject, astronomy. I proceeded to read all available books on astronomy and related subjects, which left me with the realization that many of the books were poorly written. One fact I recognized early on was the lack of the authors' methodology when presenting a theory. Instead of using the well accepted Scientific Method to present their arguments, they would simply state what they assumed was correct, with little or no factual evidence, or at least a convincing hypothesis. So many authors used this method that I refer to it as "deliberate confusion," in which they would propose a complex mathematical equation to support their ideas, such as the use of calculus—matrices, vectors, or tensors—and not providing the basic steps, normally required for a good mathematicians proof using the circular argument, such as, "If my first equation is true, then so are equations 2 and 3, none of which were proven in the first place." The end result caused me to doubt the author's knowledge of the subject. Many years later, when I attended the University of Chicago at circle campus. I was told something by my calculus professor that stuck with me all these years. It goes, "He who is convinced against his will remains unconvinced still." I learned early on that following well established guidelines, such as the Scientific Method, to state your theories and present proof thereof would tend to remove all doubt. In Chapter Eight, "The Red Shift," I use this method to present my theory of how I believe the Red Shift is accomplished, and by the way, this was the incentive that caused me to consider writing the book.

Also I would like to mention the fact that my daughter Laura gave birth to my grandson, Steven, who is the son I never had and of whom I am very proud. The culmination of this was the joy I felt watching him receive his diploma at Lake Forest College, with a double major in business and politics. In many ways he reminds me of myself when I was a young man.

Lastly I would like to dedicate this book to my wife Elaine of fifty-one years, who recently passed away. She was my one-person audience; I talked to her whenever I had ideas about things no one had ever said before. I am sure she will be watching how I do with these ideas.

INTRODUCTION

Long before recorded history, early man had wondered about the stars that appeared in the night sky. What were they? How did they stay up there? How could they come back each night in the same pattern? What caused them to twinkle? These were questions they could not answer at the time. Because of this many mystical stories and make-believe tails were concocted to explain what they thought they were looking at. Many thousands of years passed before man acquired the intellect to begin to comprehend complex problems and eventually to apply newfound methods and discoveries, helping him to better understand the natural world of wonders existing all around them. With the advent of historical record keeping, from the clay tablets and scrolls left by the Phoenicians and Babylonians to the construction of the pyramids by the Egyptians, it became evident that man's mental capacity and technical ability had improved to the point that eventually led him on the pathway for the understanding of true science. The earliest records of scientific achievements date back to the ancient Greeks. Their insight helped us to better understand how our Solar System worked. Many of their discoveries came about because of disagreements with common misconceptions and folklore. Examples include the Flat Earth Theory, which at the time was the common belief and took great effort by Aristarcus and Erothostenes to dispel their misconceptions and prove once and for all that the earth was round. Another was the misconception that the heavens rotated, not the earth, which took many years and many empirical experiments to finally realize they had it opposite. Another was the sun rotated around the earth, not the earth around the sun. I would like, at this point, to call the reader's attention to how easily an idea can be misinterpreted if no one takes the time or effort to present other possible explanations for some observations other than those of popular belief. This leads to the historical fact that the next eighteen hundred years of early astronomical history were not very productive, and during the time of the great Ptolemy, a Greek Astronomer (A.D. 100-170), some of his contributions were star charts and the "almagest," mainly for practical purposes, of navigations because of trade by ships. Unfortunately, their scientific knowledge was lacking because they did not understand the mechanics or "retrograde motion" of the planets. They chose to invent what I call one of the biggest misconceptions ever to burden astronomers, that of Epicycles, the imaginary secondary orbits, the planets scribed around and imaginary locus that existed within the normal orbit the planet in question scribed around the sun (which, by the way, was because of the viewing positions, differences between the earth and the planet in question, with respect to their orbits around the sun at a specific time). Once this was finally understood, we have not revisited Epicycles again. This misconception lasted almost fourteen hundred years and wasn't solved until the time of Copernicus and his contemporary Johannes Kepler. Although Copernicus was the motivating force in revisiting the heliocentric solar system, purposed by the ancient Greeks, he still was not able to solve the Epicycles problem, and it was left to Kepler to straighten out in 1594, fifty-one years after Copernicus' death. The Caveate here is, as I have stated before, and I repeat, unless the Scientific Method is used to explain a theory or observation and mystical unexplained reasons are given for an observation, there might well be a completely different cause for what is assumed to be obvious. I call this to the readers' attention because in Chapter 8 (The Red Shift), a parallel exists between the belief in epicycles and recession as a cause of the Red Shift. Many years passed before astronomy became an accepted science, and with the beginning of the Renaissance, referred to by some as the time of true enlightenment, there were numerous discoveries, breakthroughs, and many new theories and inventions that explained the natural physical laws that were previously misunderstood and which allowed us to progress out of the dark ages. Contributions made by just some of the great scientists of the world, like Newton, Liebnitz, Galileo, Tyco, Copernicus, and Kepler, along with many others, laid the groundwork for us to understand our natural world. I seriously doubt, without their help in advancing our knowledge and understanding, that the Industrial Revolution would ever have happened. Needless to say, there would be no United States.

 Moving ahead to more modern times, contributions were made by William Herschel in 1789, who designed and built a forty-nine-inch reflection telescope with a mirror made of steel, with which he discovered the planet Uranus. Later, in 1845, the earl of Rosse made discoveries and cataloged the spiral nebulae, and still later Ole Roemer discovered the finite speed of light (by observing the eclipse of Jupiter's moons). Gradually, over time, it became apparent that building bigger and better telescopes allowed us to compile data of significance, which helped us to better understand our universe. We are now in the theoretical development stage of constructing a gigantic telescope on the dark side of our moon, which will most certainly provide us with new data, new theories, and a greater understanding of how our universe works, along with new inventions and discoveries. Using the moon as a stepping stone will allow us to go on to Mars to repeat the exploration process, again setting up outposts on Mars and terraforming Mars, which should keep us busy for quite a long time.

These last thoughts on exploration will require new inventions and discoveries, along with funding, but I think that dollars used for this purpose will have a much better return on investment then what we have been seeing lately.

Now on to Chapter 1 - <u>Our Solar System.</u>

CHAPTER 1 - OUR SOLAR SYSTEM

THE SUN - Our Solar System, as observed today, consists of one sun, nine planets (I am electing to include Pluto for sentimental reasons), moons of the planets, asteroids, comets and bodies of lesser size. At the center of our Solar System is the sun—a star of Type G2 (on the Hertzsprung Russell Diagram, see Chapter 6), which is a member of the Milk Way Galaxy, located approximately 25,000 light years from galactic center. It take us 225×10^6 years to revolve around the center once. The sun's diameter is 864,000, miles and its density is equal to 1,409 times that of water, its mass is equal to 2×10^{27} tonnes, its volume equals 1,303,600 earth's, its surface gravity equals 27.9 times earth's, its escape velocity is 384 miles per second, its absolute magnitude is 4.83, its surface temperature is 5500 degrees C (9932 degrees F), and its core temperature is 15×10^6 degrees C (27×10^6 degrees F). Its rotational period is 25.4 days. Its diameter at the equator is 1,392,000 km (864,948 miles). Its cross section consists of a core, a radiative layer, a conductive layer, a photosphere, and a chromosphere. The

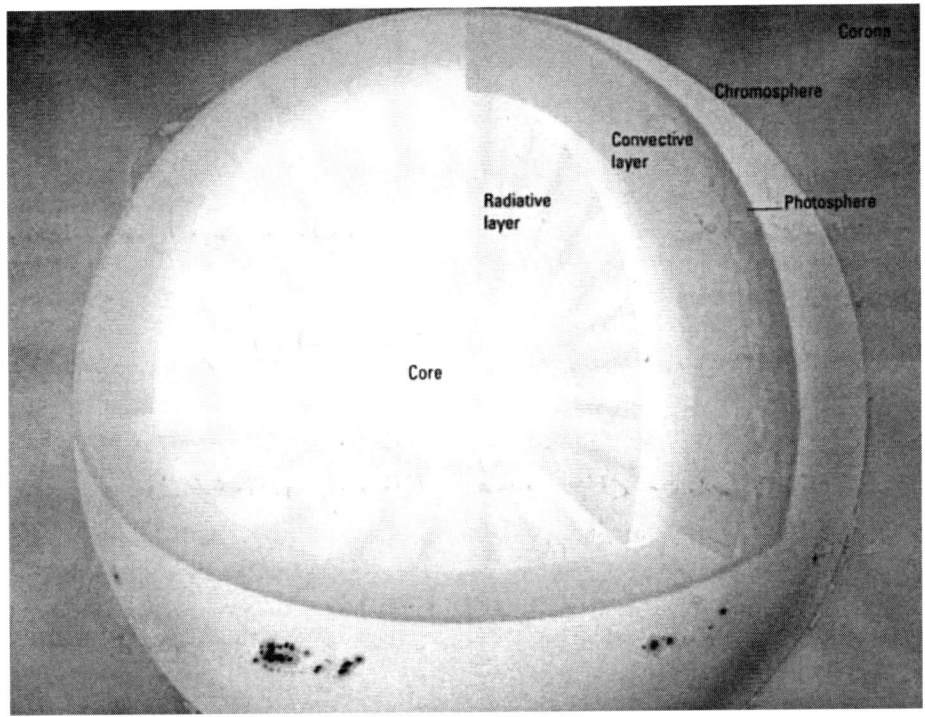

sun is located at a mean distance from earth of 92,957,130 miles; this distance is often referred to as one AU (Astronomical Unit), used by many astronomers and astrophysists to simplify their calculations and formulas when making reference to our Solar System.

The sun is a source of a broad spectrum of a light at very high intensities; therefore you should never look directly at the sun without the aid of special sun-darkening light filters, such as Welders goggles. To not use these can cause irreversible damage to your eyes.

The sun is approximately five billion years old and is composed of 70% hydrogen. The hydrogen is combined with helium at its core, which forms the atoms of helium. This process is called nuclear fusion; it takes four hydrogen atoms to make one helium atom, and this reaction is responsible for the heat and radiation the sun gives off. This process of nuclear burning is estimated to continue for another five billion years. When the hydrogen is finally consumed, the core will shrink and heat up. Then helium burning will start, the sun's outer layer will expand and cool, and at this point the sun will become a red giant, as predicted by the Hertzsprung Russell Diagram. Next it will shrink to a white dwarf, and then after all its material is exhausted to a black dwarf, this total process as described will take approximately ten billon years—long after we have gone to either another duplicated earth or to a higher plane of existence.

Lastly I would like to give to the reader a feeling for the sun's dimensions of the cross section. Starting first at the topmost layer, the photosphere measures 190 miles. Next, the convection layer is 125,000 miles, then the radiative layer is 167,310 miles, and finally, the core radius measures 140,000 miles. The stability of the sun's output was thought to fluctuate wildly initially, but after the first billion years settle down to at present, the only variations are the eleven-year sunspot cycles. Theses sunspots have a lifetime of between two hours and six months; the variations in time are believed to be caused by random subsurface conditions changing, and these are caused by thermal differences between the surface and core temperatures, along with the nuclear reaction and gravitational difference within the sun's interior. These differences are caused by inertia of mass with respect to rotation of the sun. One other anomaly I would like to point out is the sunspots only occur at latitudes of 45 degrees north and south at the start of their cycle and progress closer to, but never reaching, the equator. The maximum intensity of the sunspots is reached at approximately halfway between, or about 22 1/2 degrees, both north and south, at near 5 1/2 years after the start of a new cycle. This translates to maximum solar radiation and accounts for electromagnetic disruptions here on earth, along with other disturbances like power surges in transition lines, interference with radio and television broadcast, communication in general, and also the cause of the Aurora Borealis—and a danger to our astronauts in space. Needless to say, without the protection of the Van Allen Belt (the magnetic field around earth), and without an atmosphere, we would have been well done a long time ago.

Planetary Data

Planet	Distance from the sun in miles x 10^6	Diameters in Miles x 10^6	Sidereal Period In years or days	Mass in Tonnes x 10^{18}
Mercury	36	3,031	88 d	360
Venus	67	7,521	224.7 d	5360
Earth	93	7,926	365.3 d	6580
Mars	142	4,221	1.9 y	705
Jupiter	483	89,405	11.9 y	2.09 x 10^6
Saturn	886	74,897	29.5 y	625 x 10^3
Uranus	1,783	31,763	84 y	96 x 10^3
Neptune	2,794	31,402	164.8 y	116 x 10^3
Pluto	3,666	1,444	248.4 y	1.1 x 10^3

Note: To convert miles to km, multiply by 1.609344. I have provided the above data to give the reader a table to go to for a quick glance to compare data of the individual planets.

Solar Data - Sun

Distance from Earth	149,597,893 km (92,955,821 miles)
Mean Distance from Center of Galaxy	25,000 light years
Velocity around Center of Galaxy	220 km/s (136.7 miles/s)
Revolution Period around center of Galaxy	225,000,000 years
Density, water =1	1409
Mass, earth = 1	332,946
Mass	2×10^{27} Tonnes
Volume Earth = 1	1,303,000
Surface Gravity of earth = 1	27.9
Escape Velocity	617.5 km/s (384 Miles)
Spectrum	G2
Surface Temperature	5500° C
Core Temperature	15,000,000° C
Rotation Period	25.4 Days
Diameter	1,392,000 km (865,000 miles)

UNDERSTANDING THE UNIVERSE

Mercury

Planetary Data

Sidereal Period	87.97 days
Rotation Period	58.646 days
Orbital Velocity	47.87 km/s = (29.76 miles/s)
Orbital Inclination	7° 00' 15.5 " Axial Inclination 2°
Orbital Eccentricity	0.206
Diameter Equator	4878 km – (3030 miles)
Distance from the Sun (mean)	57.9 x 10^6 km (35,987,320.53 miles)
Density, Water =1	5.5
Mass, Earth = 1	0.055
Volume, Earth = 1	0.056
Surface Temperature	Dark side – 170°C Sun side + 350° C

MERCURY - The first planet in orbit, moving away from our sun, is Mercury. Its orbit has a mean distance of 57,900,000 km (35,987,320 miles). It takes 88 days to make one revolution around the sun and rotates on its axis once every 58 days, which is unique compared to the other planets. It orbital velocity is 29.76 miles/sec, its axel inclinations is 11°, its diameter is 3030 miles, its density is 5.5 times that of water, and it has a volume of 0.056 times that of Earth. Its escape velocity is 2.7 miles/sec, its surface gravity is 0.38 that of earth's. Its surface temperature is + 350° C (662° F) on the sunlit side and -170° (-274° F) on the dark side, and its magnitude is (-1.9).

Because of its proximity to the sun, it is almost impossible to get images of its surface as viewed from earth. This is because of practically no separation between the sun and Mercury. Our Mariner 10 space probe sent beautiful high resolution images of Mercury's surface showing craters two and three times larger than those of

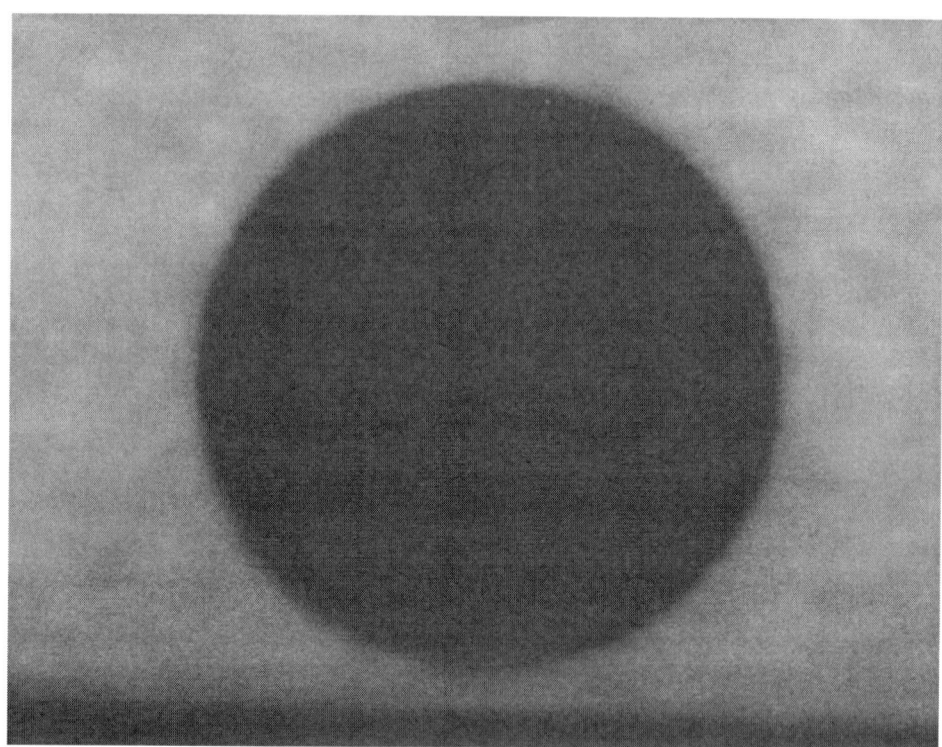

earth's moon, some over 350 miles across. This makes sense if you consider Mercury's proximity to the sun, being in a direct path of asteroids being sucked into the sun by its enormous gravity. The Mariner 10 also revealed other unique properties such as the fact that Mercury's atmosphere was only 1/10,000,000,000 that of earth's–probably stripped away by the sun's gravity. A magnetic field was detected, although it was only 1% of earth's and had two poles (north and south). I don't think anyone has found a magnetic monopole yet! Mercury is also the second-most dense planet, second only to earth, having an iron-rich core equal to 70% of its mass, with the other 30% being rocky. The core is thought to be molten generating the magnetic field. Because of its eccentric orbital variations, many scenarios have been written, one of which I will relate. If an observer were to be on Mercury at a certain time, his observations would go something like this. The sun would rise overhead, stop, then go backwards for eight days, and then resume its original motion. These strange actions would be caused by Mercury's revolution period, orbital period, and eccentricity of orbit. The Mariner 10 probe took hundreds of images, showing craters with central peaks so high they could only have been caused by impacts into a semi-plastic surface. The time of these impacts was estimated to be between three and four billion years ago, approximately the same time as the impact craters on our moon. I wonder what great cataclysm happened then. Could it have been a collision with another galaxy? Would it be possible to work backwards, running the heavens in reverse like some planetariums do and see what happened three to four billion years ago? Later, in discussing some of our other planets, I will revisit this same phenomenon and try to piece together what happened.

Robert E. Smolinski

Venus

Planetary Data

Sidereal Period	224.701 Days
Rotation Period	243.16 days
Orbital Velocity	35.02 km/s (21.76 miles/s)
Orbital Inclination	3° 23' 39".8 / Axial Inclination 178°
Orbital Excentricity	0.007
Diameter Equation	12,104 km (7543 miles)
Distance from sun (mean)	108.2×10^6 km (67,232,363 miles)
Density, water =1	5.25
Mass , Water =1	0.815
Volume, Earth =1	0.86
Temperature (surface)	+480 °C (+896 ° F)

VENUS

The next planet we come to as we move away from the sun is Venus, at a mean distance of 108,200,200 km (67,232,487 miles). It takes 224.7 days to go around the sun once and rotates on its axis in 243.16 days. (It's interesting that it takes longer to rotate once on its axis than it takes to go all the way around the sun). This is referred to as retrograde motion; not only is it weird, but it also rotates backwards—east to west. Why? How could this be possible; it would defy all laws of physics if it were a natural action, but if what we see is what we get, we can then assume that this strange behavior is not natural and therefore must have been caused by some interaction with another body. Envision two possibilities: 1.) Venus was grazed by another body of considerable mass, as it was pulled into the sun, or 2.) (This is the one I like best.) Venus does not belong in our Solar System at all; it is a visitor from elsewhere in space, entering into our

UNDERSTANDING THE UNIVERSE

Solar System sometime in the past, perhaps one to three billion years ago, causing great disruptions on its way in. Planets that show evidence of catastrophic disturbances are Uranus, knocked 98 degrees off its axis, the asteroid belt, where only pieces of what we referred to as meteoroids and asteroids now occupy the orbital zone that once could have been where one or more planets existed, but because of collision and rebounding impacts were reduced to what we see today—billions of pieces of debris, some as large as Ceres— 560 miles in diameter and others 20 - 150 miles in diameter, with the greatest number, perhaps trillions, which include sizes down to that of a grain of sand. Also, Mars shows evidence of abnormal aging, the Great Scar, and two misshapen moons, which I will go into later, when I talk about Mars' anomalies. How many times has the reader been presented with scenarios describing how our moon came about? I can remember at least three good ones, but I don't ever recall pointing the finger at Venus being the culprit. I will suggest at this point what I envision happened. As Venus made its way toward the sun, its trajectory was such that it collided with earth, a glancing blow knocking off sufficient mass. This eventually became our moon, then it continued its deflected trajectory to take up an elliptical eccentric orbit between earth and Mercury; this orbit and

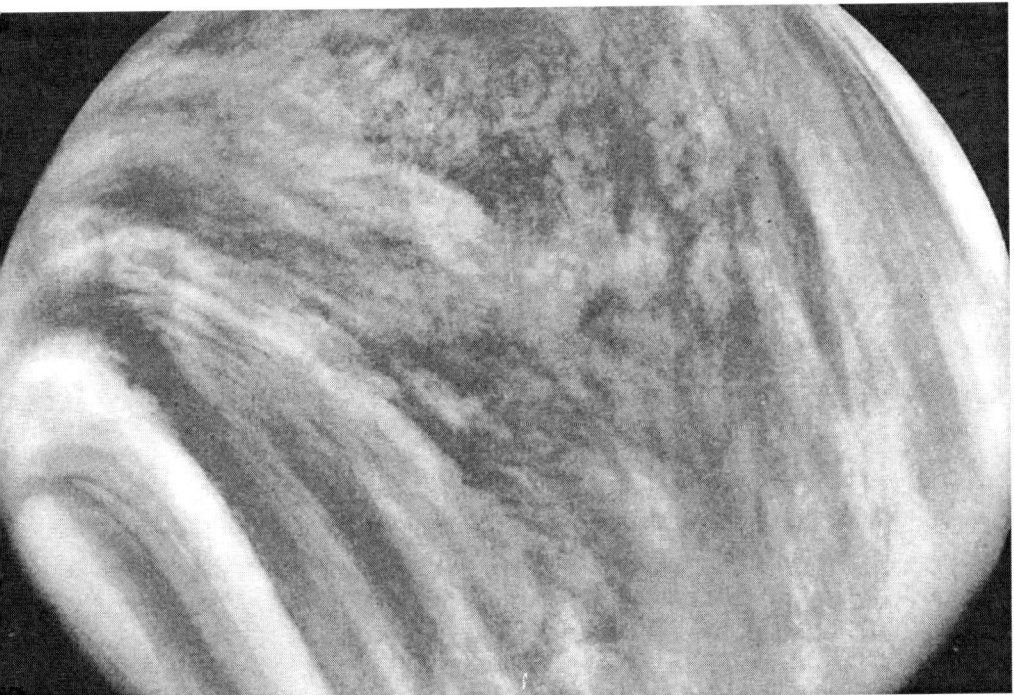

that of earth's eventually became more circular because of gravitational moments of inertia, and they settled down to what we observe today. Other things about Venus that don't fit the puzzle are its temperature, almost 900° F—hot enough to melt lead. This explains why the Russian probes sent there didn't last very long. The corrosive acid atmosphere, at a density 89 times that of earth's, caused failure in a matter of minutes. The Magellan probe sent to Venus took many images of the surface of Venus, some of which showed what we would like to call mountains, but were 16 miles in diameter and 2400 feet in height. What mechanisms originating in our Solar System could be responsible for structures such as these, or would this be considered to be a normal development for a body from elsewhere? Venus's dynamic parameters are as follows: Its orbital velocity is 35.02 kms/s (21.76 miles/s). Its orbital inclinations are 3º 23'39.8". Its axial inclination is 178º (what?), its orbital eccentricity is 0.007, and its equatorial diameter is 12,104 km (7523 miles). Its density is 5.25 times that of water. Its mass - water = 1 - .815. Its volume is 0.86 - earth = 1 and its gravity is 0.903 - earth = 1. One last additional fact: Nowhere on the surface of Venus could any impact craters be found. Could it be possible that Venus was not here in our Solar System, when all the other planets and satellites were bombarded, evidenced by the fact we see craters, or evidence thereof, on all the moons of the planet in our Solar System, as well as on some planets?

Earth

Planetary Data

Sidereal Period	365.3 days = 1 year
Rotation Period	23 hr – 56 m -04 s
Orbital Velocity	67,000 mph (107,826 kmph)
Orbital Inclination	0 ° / Axial inclination 23.4°
Orbital excentricity	0.017
Diameter Equation	12.756 km
Distance from Sun (mean)	149.6 × 10 ^6 km 9,295,713,036 miles)
Density, Water = 1	5.52
Mass, Earth =1	1
Volume, Earth = 1	1
Temperature (Surface)	22 ° (71.6 ° F)

EARTH

I could write a book just on the planet earth, primarily because we have billions of times more data than any other planet or stellar body, but my topic is the universe, and for the sake of brevity, I will describe only the vital solar planetary data. To continue, our planet earth is third in orbit from our sun at 149,000,000 km (9,295,713,036 miles). Our year is 365.3 days, the time it takes earth to go around the sun once. Its rotational period is 23 hrs, 56 min, 04 s = 1 day. Its axial inclination is 23.5 º. Its escape velocity is 6.952 miles/s. Earth's density is 5.52 times that of water, and the atmospheric pressure is 14.7 PSI at sea level. Its gravity = 1; its surface temperature (ave) is 71.6º. Its diameter at the equator is 7,932.83 miles. Earth's surface is 78% water. Earth's cross sections are unique, primarily because it's the only planet in our System that has liquid water; this in turn causes the upper crust to have a capillary motion. This motion is referred to as tectonic plate motion. Below the crust we have a mantel, a transitions zone, a lower mantel, an outer molten core, and finally an inner liquid molten iron core which, because of its rotational differences with adjacent layers, is responsible for generating earth's magnetic field, and also our volcanoes, earthquakes, and the force behind the movement of plate tectonics. This, by the way, is also the reason we have pure elements created far below the earth's surface, where the temperature and pressure necessary to form pure elements are present. Some of these precious elements are diamonds, gold, silver, and most of the elements of the Periodic Table. The earth's atmospheric pressure varies from 14.7 PSI at sea level to "0" PSI at 300 miles above the surface. The atmosphere is composed of 78% nitrogen, 21% oxygen, and 1% trace elements. The temperature of the atmosphere varies in an unexpected way and does not change linearly as a function of altitude, as might be expected—for example, taking a cross section at the equator. Starting at the surface, the temperature (ave) is 22° C (71.6° F), dropping to -44°C at 10 miles at the top of the first layer (the troposphere), then gradually rising to +15 °C at the top of the second layer (the stratosphere), then it starts dropping and continues to drop as you go out in space.

Earth

Earth has one moon, which was thought to be created by a collision (see Venus and Chapter 6). Our moon orbits at a mean distance of 238,828 miles (and is increasing 1.5 inches per year). It orbits earth in 27.321 days; it has an orbital velocity around earth of 2286 miles/hr. Its diameter is 2160 miles; its orbital inclinations are 5° - 9″. With respect to earth, equatorial plane is thought to be responsible for variations in the yearly weather pattern of earth, primarily because of the positional differences of the moon with respect to the latitude and longitude interactions with respect to time. The moon's density is 3.34 times that of water; its mass is 0.012 that of earth's, and its volume is 0.020 that of earth's. Its gravity is 0.165 that of earth's, and its escape velocity is 1.48 miles/s. As mentioned above, the moon's gravitational effect not only affects the weather but also the ocean tides around the world, I believe the gravitational effects of the moon, as well as the sun and planets, are responsible for 80% of our earth's weather patterns, including global warming. The other 20% is caused by excessive amounts of CO_2 and the cutting and burning of our forests, the CO_2 being generated by burning of fossil fuels, gasoline, jet planes' exhaust (including vapor trails), power-generating plants, steel mills, and many other sources of CO_2 pumped into the atmosphere. We have to stop this craziness and start using solar photovoltaic panels and geothermal wind and wave power to generate our power needs and to eliminate our dependence on petroleum and fossil fuels. Also, forget nuclear power plants because they are dangerous (look what happened to Chernobyl). What do you do with the spent rods? No one wants them stored in their state; would you? Let's not even talk about breeder reactors; they are the most dangerous of all, primarily because they can be used to manufacture fuel for nuclear bombs. Look at Iran for an example. Returning to the subject, the surface of the moon shows many impact craters caused by collisions eons ago, and because the moon has no atmosphere to erode them, they appear much as they did when they were created. The craters are very well documented and are named after many of our great astronomers and scientists. Just to name a few, there are Copernicus, Erothoteness, Kepler, Tyco, and many others. Beside the craters there are large dark areas called Marias (Latin for Seas); they are also well documented and marked with exact latitude and longitude and are believed to have been caused by upwelling of lava flows when the moon was first forming.

The United States is the only country that has landed astronauts on the moon (six times). They conducted scientific experiments and brought back surface samples for further laboratory analysis. Watching Neil Armstrong step out on the moon and make his immoral statement, "That's one step for man and one giant leap for mankind," was one of the proudest moments of my life. I was glad to see them come home safely. We will begin going back very soon to set up permanent outposts and much larger telescopes on the dark side for better viewing, and hopefully we will acquire new data of distant galaxies and quasars and improve our knowledge and understanding of how our universe truly works, using the moon as a stepping stone for eventual manned space flight to Mars and beyond. (For a look to the future, see Chapter 12).

EARTH'S MOON

Satellite Data

Distance from Planet	384,4000 km (238,828 miles) mean
Orbital Period	27.321661 days
Orbital velocity	3680 km/h (2286 Miles/h)
Orbital Inclination	5 ° - 9" (18 year cycle)
Orbital Eccentricity	Apogee 252,681 Miles – Perigee 221,438 miles
Diameter Equator	3476.6 km (2160 miles)
Density, Water = 1	3.34
Escape Velocity	2.38 km/s (1.48 miles/s)
Mass, Earth = 1	0.012
Volume, Earth = 1	0.010
Temperature	-240 ° F to +240 ° F = -150 ° C to +120 ° C

Robert E. Smolinski

Mars

Planetary Data

Sidereal Period	686.980 days
Rotation Period	24 hr – 37 m – 22.6 s
Orbital Velocity	24.1 km/s – (15 miles/s)
Orbital Inclination	1 ° 50' 59.4'' / Axial Inclination 24.0 °
Orbital Excentricity	0.093
Diameter Equator	6794 km (4222 miles)
Distance from Sun (mean)	227.9 × 10^6 km (141,610,949.7 miles)
Density, Water = 1	3.94
Mass, Earth = 1	0.107
Volume, Earth = 1	0.150
Temperature	-23 ° C (-9.4 ° F)

MARS

Next stop: planet number four, Mars, located between earth and the Asteroid Belt. Its data is very interesting because it's been the source of examination for hundreds of years by many famous astronomers, and it's interesting to compare the old and new data to see our progress. Mars is located at a mean distance of 227,900,000 km (141,610,494 miles) from our sun. It takes 686.980 days to complete one orbit and 24 h - 37 m - 22.6 s to rotate once on its axis (one day). Its orbital velocity is 15 miles/s, its orbital inclination is 1 º 50' 59", its diameter is 6794 km (4222 miles) at the equator, its density is 3.94 times that of water, its mass is 0.107 that of earth's, its volume is 0.150 that of earth's. Its escape velocity is 3.1 miles/s, its surface gravity is 0.380 that of earth's, and its (ave) surface temperature is -23° C (-9.4 ° F).

UNDERSTANDING THE UNIVERSE

In 1659 Dutch astronomer Christen Huygens studied the surface of Mars using his newly designed telescope and noted dark and light areas he believed to be seas and land. Later, in 1877, G.V. Schaperilli drew lines he thought he saw in his telescope and referred to them as "canali," which in Italian means channel, but were mistaken to mean "canals," leading to the theory that Mars at one time in the past had water and supported intelligent life. We now know that the lines he thought he saw were in reality rifts and valleys carved out by as yet unknown forces, but the most recent popular belief is that water was the principal force.

Both the United States and Russia have sent a number of space probes to the Martian surface, sending back many images and data for study. Our Mariner 9 in 1971 sent over 7000 high-resolution images of the Martian surface and provided reams of data not previously acquired. Later, our Viking Lander and Rover provided realistic panoramic images of the actual surface features, including rocks and boulders of sizes three meters on down. The images and soil samples were extremely valuable data, although there was no evidence of water or life.

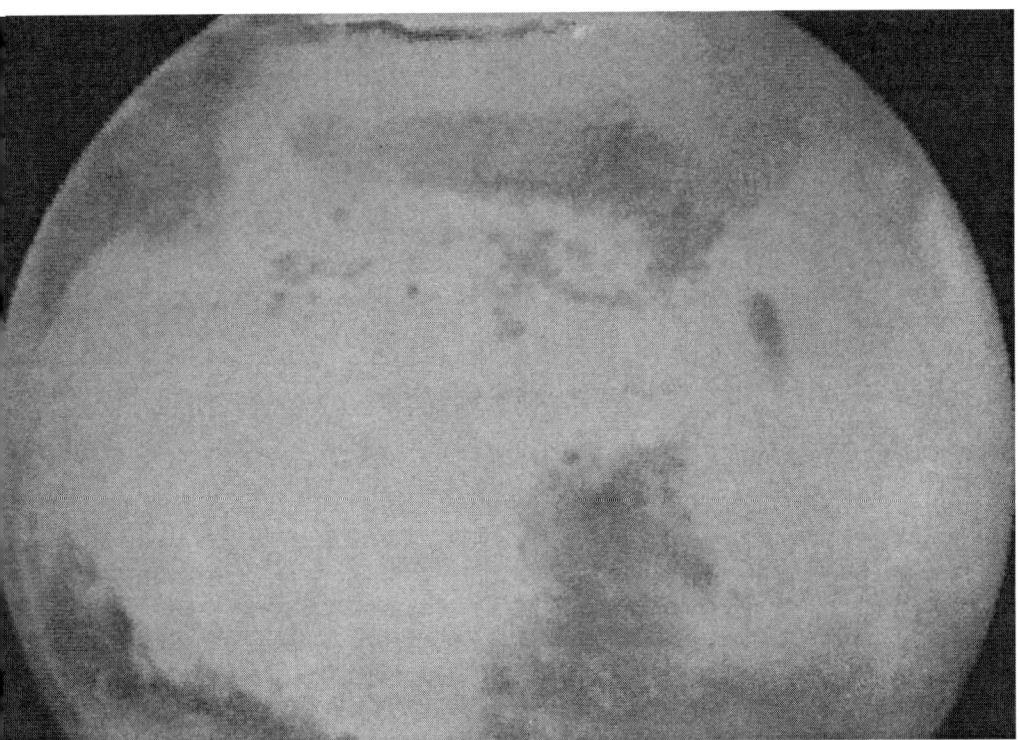

There are a number of major surface features on Mars, the most prominent of which is the Great Scar (Noctis Labyrinthus - Vallies Marineris) streaking diagonally across the Martian surface for 3000 miles, the depth and width of which is greater than our Grand Canyon. What could have caused this? A group of major surface features is the Olympus Mons—Ascranes Mons and five other lesser volcanic chimneys located on each side of the great Scar, along with two gigantic surface bulges, Tharsiss and Elysium, suggesting they were all caused by extreme volcanic activity. Many of the dormant volcanic chimneys showed evidence of outgassing. How could this be possible if Mars had no molten liquid core or water? Because of this evidence, I suggest that Mars at one time in the past did have water and a molten core. If this is the case, then what happened to Mars to cause such destruction on such a large scale? Another unusual feature is the presence of two satellites, Phobos and Deimos, which have diameters of 17 miles and 9 miles respectively and orbit at 5800 miles and 14,500 miles. Where did they come from? Could they have come from the interior of Mars? Obviously they have been there a long time, evidenced by their pockmarked cratered surfaces. Some of the craters are as large as 3 miles in diameter. What cataclysm could have caused these major anomalies, and are they in any way related? (See chapter 6 for one explanation.)

On September 18 and October 15 of 1996, our Hubble telescope took two high-resolution images of Mars, revealing macro features of the ice-covered poles and dust storms. The polar ice caps were believed to be composed of a thin layer of frozen CO_2, accounting for changes in shape throughout the Martian season. In 1997, the Viking probes 1 and 2 and Pathfinder recorded surface temperatures of -13° C (8.6° F) daytime and -75° C (-103° F) at night. These temperatures are well below the freezing point of water, suggesting life as we know it is not possible.

Comparing our new data with the old allowed us to dispel many of the fantasies and tails about Mars which, at their conception, had no basis in good science anyway.

UNDERSTANDING THE UNIVERSE

Phobos

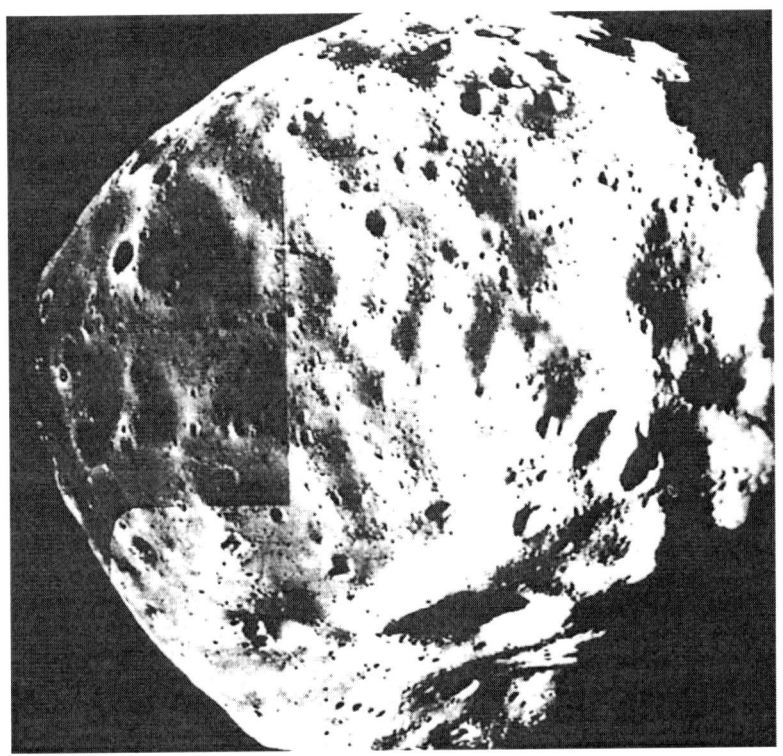

Satellite Data

Distance from Planet	9270 km (5760 Miles)
Orbital Period	7 h – 39 m
Orbital Velocity	1 orbit – 7 hrs – 39 mins
Orbital Excentricity	Near "0"
Diameter Equator	27 km (17 miles)

PHOBOS

Phobos is the larger of two moons orbiting Mars. It is 17 miles in diameter (27 km) and orbits at 9300 km (5800 miles). It is not a perfect sphere and shows pockmarks from crater impacts. It was discovered by Asaph Hall in 1877.

Deimos

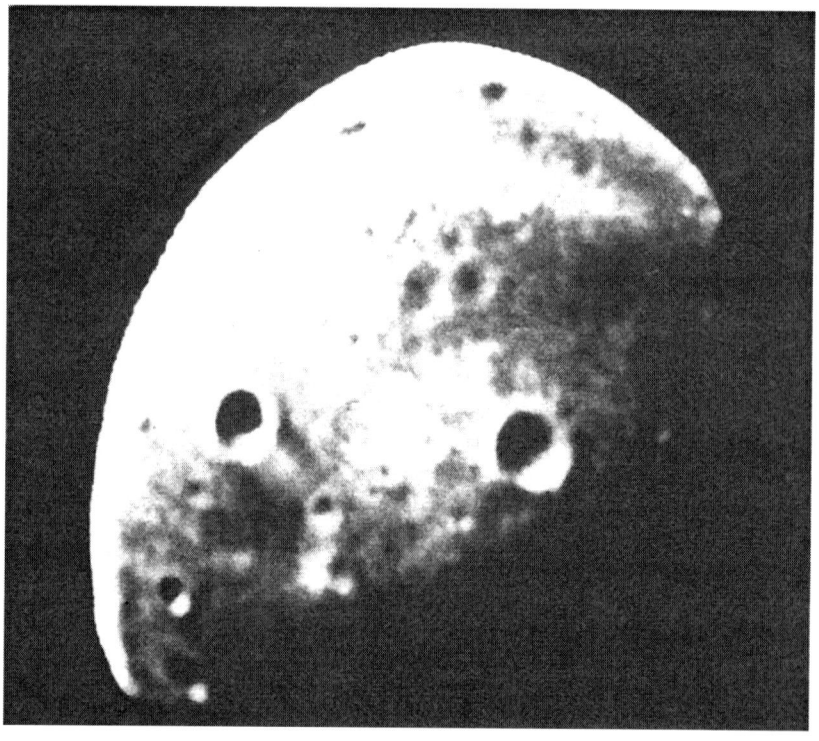

Satellite Data

Distance from Planet	23,500 km (14,602 miles)
Orbital Period	30 hrs -18 mins
Orbital Velocity	1 Orbit = 30 hrs 18 mins
Orbital Excentricity	Near "0"
Diameter Equator	15 km (9 miles)

Deimos is one of two moons' orbiting around Mars. It is the smaller of the two moons with a diameter of 10 × 15 km. It orbits around Mars at 23,500 km. Little is known about Deimos because it is so small. Deimos was discovered in 1877 by Asaph Hall.

UNDERSTANDING THE UNIVERSE

Asteroid Belt

In 1800 Franz Van Zach discovered the first asteroids after looking for what he thought were missing planets, along with Giuseppe Piazzi, who discovered Ceres in 1801.

THE ASTEROID BELT
Moving still farther away from our sun, between the orbits of Mars and Jupiter, at a distance of 228 - 778 million km, there exists what

I like to refer to as the remnants of where two or more planets originally existed when our Solar System condensed. What is left is what we see today. The asteroids vary in size, the largest being Ceres at 940 km in diameter and Pallas at 580 km. The next twenty are lesser in size; the smallest of the group is Ida at 52 km. These are the major remnants; the balance of the 100 billion pieces of debris vary in size from Idas, 52 km, down to a grain of sand. What makes the Asteroid Belt unique is its location and composition. Its location is exactly where planets would have coalesced from the accretion disc, as did the other planets. Their compositions were examined by a spectroscope and found to consist of all the elements (chemical, mineral, and organic) that would be found on a planet or planets. It's a foregone conclusion that these constituent parts could have never developed on such comparatively small bodies and therefore leaves us with only one conclusion: These pieces (the Asteroid Belt) must have belonged to a much larger body or bodies, and this suggests that some sort of disaster took place. A collision of such dimensions, it rocked the entire Solar System. One other point I think worth mentioning in this theoretical collision: I wonder if any astronomer or scientist ever estimated the total mass of all asteroids added together, and using this figure to estimate, if it were possible, this mass could equal that of one or two planets occupying and orbiting between Mars and Jupiter. This information, once established, would reinforce a hypothesis to explain not only the presence of the asteroids but also the intrusion of a visitor into our Solar System from elsewhere in space. A last point I would like to make is J.E. Bode's empirical rule of numerical progression. Although never proven, it suggests that the physics of our solar system would substantiate the presence of one or more bodies in numerical order where only the Asteroid Belt now exists.

In 1766 J.D. Titus, and later in 1768 J.E. Bode, presented an argument for the existence of other planets located where the orbits of the asteroids are now, between 141×10^6 miles and 483×10^6 miles away.

UNDERSTANDING THE UNIVERSE

Their empirical rule as they presented it was:

4	4	4	4	4	4	4	4	4	4	4
	+3	3x2	3x2^2	3x2^3	3x2^4	3x2^5	3x2^6	3x2^7	3x2^8
4	7	10	16	28	52	100	196	388	772
36	67.2	92.9	141.6		483.4	886	1782	2792.9	3671.9
Mer-cury	Venus	Earth	Mars	Asteroids *	Jupiter	Saturn	Uranus	Neptune	Pluto

*This is the area that suggests something is missing. It might be interesting to revisit Bode's rule of numerical progression once again, and we should be mindful of the fact that the present orbits or the planets might not have been exactly where they were three billion years ago.

Jupiter

Planetary Data

Sidereal Period	4332.59 days – 11.9 Earth years
Rotation Period	9hrs – 55m – 30s
Orbital Velocity	13.06 km/s (8.1 miles/s)
Orbital Inclination	1° 18' 15.8" / Axial Inclination 3.0°
Orbital Excentricity	0.048
Diameter Equator	143,884 km (89,424 miles)
Distance from Sun (mean)	778 x 10^6 km (483,560,196.4 miles)
Density, Water = 1	1.33
Mass, Earth = 1	317.89
Volume, Earth = 1	1318.7
Gravity, Earth = 1	2.64

JUPITER

The next planet we come to as we move away from our sun is planet number five, Jupiter, at a mean distance of 778,000,000 km (483,530,143 miles). It is by far the largest planet and almost has sufficient mass to start nuclear burning on its own. To get a feeling of its gigantic size, over eleven of earth's diameters placed end to end would equal one of Jupiter's diameter. Although its physical size is so large, its density is only 1.33 times that of water, which means its total volume is composed of mostly gas, with a rocky or metallic surface far below. Its orbital period around the sun is 11.86 years, and it rotates on its axis in 9 h - 55m - 3 s. Its orbit is inclined by 1.3°, and its axial inclination is 3°. Its gravity is 2.64 times that of earth's, and its escape velocity is 60.22 km/s. Its volume is 1319 times that of earth's, and its mass is 317.9 times that of earth's. Its surface temperature is -150°C, and its diameter is 143,884 km at the equator and 133,700 km across its poles, for a difference of 10,184 km. This difference is caused primarily by its fast rotation and its density. For comparison, our moon has a diameter of 3475 km, three times less than Jupiter's polar diameter difference. Because of Jupiter's size it is easily viewed from earth, and because this has been the source of much investigation for the last 500 years or so, this allows us to compare data taken today with that taken long ago, revealing many dynamic inconsistencies such as orbital period, rotational period, and orbital eccentricity. Comparing this data, we can then make better estimates of how Jupiter affects our Solar System. Jupiter has many distinguishing features such as The Great Red Spot, which I have cataloged on many occasions, with only a conservative 600 power refractor. Additional prominent features are its bands of color, representing cloud formations of different elements at different latitudes and being accelerated amazingly in opposite directions at tremendous velocities. Thousands of images of Jupiter have been taken by Voyager, Cassini, and Hubble, revealing great detail and, in some instances, the recording of special events, like the collisions with the 13-25 pieces of the

Shoemaker - Levy -Comet. Jupiter has 60 moons at last count; of these, four are giant satellites. The largest of these is Ganymede, with a diameter of 5286 km (3266 miles). Next is Callisto at 4806 km (2986.9 miles), then Io at 3637 km (2260 miles), and, the last of the big ones, Europa at 3130 km (1959 miles). The other 56 moons are much smaller in diameter, the smallest of which is LEDA at 10 km (6.215 miles). The major moons are very easy to resolve; all that's needed is a pair of binoculars or a low-power telescope, but in order to appreciate their unique differences, it's necessary to see the images taken by Hubble or our space probes. I will describe them in order of their distance in orbit from Jupiter. First is Io, an extremely active moon. Volcanos can be seen ejecting material hundreds of kilometers above its surface. Its crust is in a plastic state, like a pot of molten sulfur. Recorded temperature extremes show differences of +500° C for some volcanos and -150°C for flat, less active areas. Europa is next; in contrast to Io it appears as an inert ice-coated moon, with expansion cracks covering its surface, and beneath the ice is thought to exist vast oceans of liquid water that perhaps contains early life. This is a popular theory and is on the drawing board for future exploration. Next is Ganymede, the third moon from Jupiter and by far the largest. It has both icy and rocky patches showing on its surface, which is larger than the planet Mercury and was first seen by the Chinese observer Gam De in 364 BC. Next, after Ganymede, is Callisto, the second largest moon, the outermost major moon thought to have a metallic core, surrounded by a mantel of water and ice mix and an icy surface crust. There are no signs of surface tectonic activity, and it appears totally inert. The 56 remaining smaller moons are arranged mostly inside the orbit of the major moons, and many outside. They do exist, but little is known about their physical nature, except orbital and basic physical data, primarily because they are so small and hard to detect.

Io

Satellite Data

Distance from Planet	421,600 km (261,800 miles)
Orbital Period	1.769 days
Orbital Velocity	42 hrs
Orbital Inclination	0.04°
Orbital Excentricity	0
Diameter Equator	3637 km (2259.65 miles)
Density, Water = 1	3.55
Escape Velocity	2.56 km/s
Mass, Earth = 1	=0.02
Volume, Earth = 1	~0.038

Io

Io is larger than Earth's moon by 100 miles and orbits Jupiter at 262,000 miles (421,600 km). Io is volcanic and has a plastic surface. Because of its proximity to Jupiter's gigantic gravitational field, it is subjected to tremendous internal frictional forces which generate tremendous heating and are responsible for its volcanism. The internal temperature can reach values of over 2000° F (1093° C). This difference of temperature from the surface causes the most violent conditions for any body in our Solar System. It is the first in orbit around Jupiter and was first seen by Galileo in 1610, using the very first astronomical telescope. It has been noted that another astronomer named Simon Maries could have also been credited with the sighting earlier.

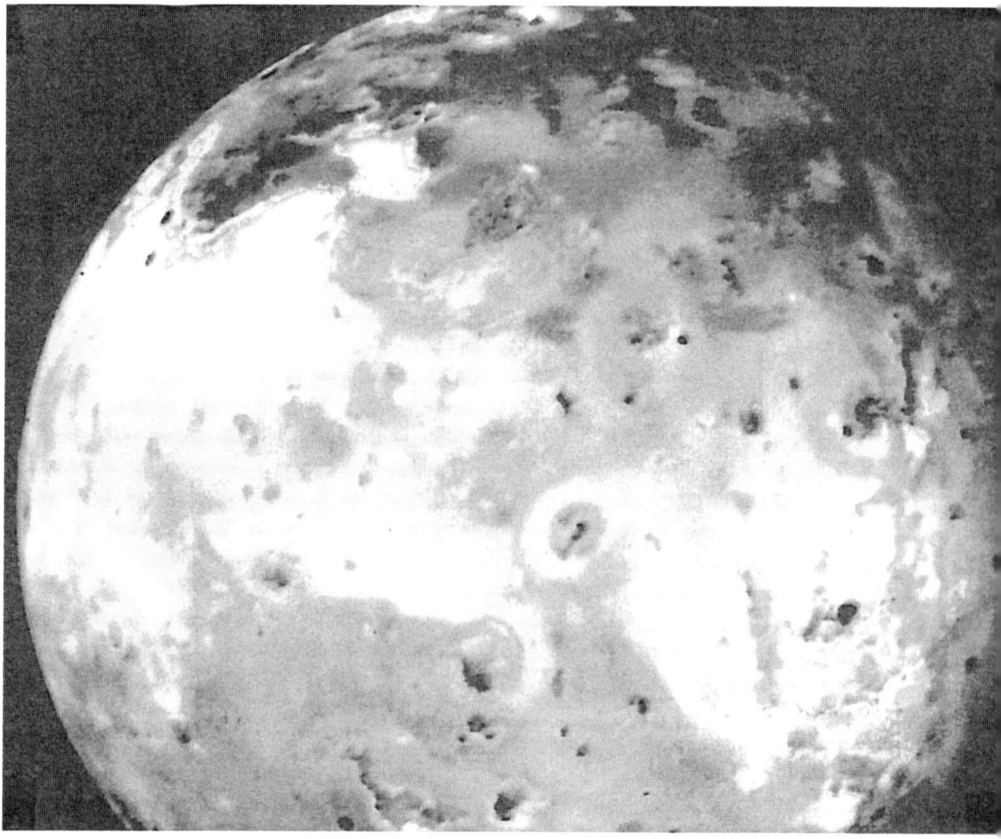

Europa

Satellite Data

Distance from Planet	670,900 km (416,878 miles)
Orbital Period	3.551 days
Orbital Velocity	
Orbital Inclination	0.47
Orbital Excentricity	0.009
Diameter Equator	3122 km (1939 miles)
Density, Water = 1	3.04
Escape Velocity	2.10
Mass, Earth = 1	
Volume, Earth = 1	

EUROPA

In 2012 the United States will send a space probe to Europa to analyze the surface, to determine if 1.) the surface ice is water ice, 2.) beneath the surface is liquid water, and 3.) the subsurface liquid water contains any life at all, microscopic or otherwise. It was first seen by Galileo Galilei in 1610 and has a very smooth surface devoid of almost all craters. The German Simon Maricus (1573 - 1624) is thought to have seen Europa as well. Who was first remains to be seen.

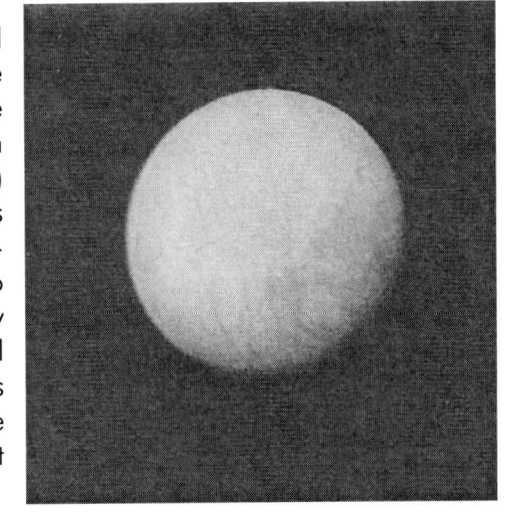

Ganymede

Salellite Data

Distance from Planet	1,070,000 km (664,867 miles)
Orbital Period	7.155
Orbital Velocity	
Orbital Inclination	0.21
Orbital Excentricity	0.002
Diameter Equator	5268 km (3274 miles)
Density, Water = 1	1.93
Escape Velocity	2.78
Mass, Earth = 1	
Volume, Earth = 1	

GANYMEDE

The largest moon in our Solar System, larger than Mercury and Pluto with a diameter of 3268 miles, Ganymede approaches 0.78 times the size of Mars. Its surface is a mixture of rock and water ice, along with many different elements and chemical gases in a frozen state. It has an iron core and a fairly strong magnetic field. Its surface has vast areas of dark and light patches, photographed by the Galileo, Space Probe, and the Hubble Telescope. It exhibits evidence of tectonic movement and has craters of a variety of sizes. The surface is marked with ridges and valleys and was first sighted by Galileo Galilei and Simon Marius in 1610.

Callisto

Satellite Data

Distance from Planet	1,880,000 km (1,168,177 miles)
Orbital Period	16.689 days
Orbital Velocity	
Orbital Inclination	0.51°
Orbital Excentricity	0.007
Diameter Equator	4806 km (2986 miles)
Density, Water = 1	1.81
Escape Velocity	2.43
Mass, Earth = 1	
Volume, Earth = 1	

CALLISTO

The second-largest moon of Jupiter, at approximately 3000 miles, Callisto has a surface composed of a mixture of rock and water ice, has no detectable magnetic field, and shows many impact craters, which expose icy patches from below. Callisto is the most distant of the major four moons and shows no sign of any surface activity, although at last count there were a total of 60 moons, all of which are considerably smaller than the four major moons and located at various distances from planet Jupiter, sighted in 1610 by Galileo Galilei and Simon Marius.

Saturn

Planetary Data

Sidereal Period	10,759.20 days – (29.5 Earth Years)
Rotation Period	10 hrs – 13 m – 59 S
Orbital Velocity	9.6 km/s (6.0 miles/s)
Orbital Inclination	2° 29' 21.6" / Axial Inclination 26.4°
Orbital Excentricity	0.056
Diameter Equator	120,536 km (74,914 miles)
Distance from Sun (mean)	142x10^6 km (886,941,388.5 miles)
Density, Water = 1	0.71
Mass, Earth = 1	95.17
Volume, Earth = 1	744
Gravity, Earth = 1	1.16

SATURN

Planet number six in orbit around our sun is Saturn, the only planet that has rings around it visible from Earth and can easily be resolved with binoculars or a low-power telescope. The rings consist of a variety of materials, ranging in size from a grain of sand to a small house. The rings were first seen by Christian Huygens (1629-1695), and we know today they are arranged in groups and divided into segments, which have been given the designations A, B, C, D, E, F, and G. The Cassini/Huygens space probe recently visited Saturn and took beautiful, high-resolution color images while passing through the rings, and did so without a collision with any of the ring materials, which, by the way, was quite a feat in itself. It also provided planetary images and data never before seen. The Huygens probe was then sent to the surface of Titan, Saturn's largest moon, which is the only satellite to have an atmosphere. It parachuted to a soft landing and proceeded to transmit megabytes of data, revealing an atmosphere of nitrogen and methane, with liquid methane on the surface. Cryogenic volcanos abounded, reminiscent of what early earth might have looked like billions of years ago, but of course with completely different elemental thermal ambient conditions. The atmospheric pressure on Titan is 1.5 times that of earth's and has a temperature of -168° C, very close to the triple point of methane. (The triple point, just in case you didn't know, is the point at which methane can exist as a solid, liquid, or gas.) With methane oceans, the liquid could boil off to explain the atmosphere as proposed by Carl Sagan. The Huygens Probe transmitted data for only 41 minutes before succumbing to the extremely cold temperature, but in that time it transmitted data of millions of bytes that will be analyzed here on earth for many months. Titan holds most of the ingredients required for life to exist, and scenarios of the future suggest that billions of years from now, when our sun swells to a Red Giant, maybe Titan could be a surrogate home, but that's one optimistic view. Because of Titan's low escape velocity, it's most likely that whatever atmosphere existed at that time would be lost to space.

Saturn's distance from our sun is 142,000,000 km (886,941,388 miles) and takes 29.47 years to make one tour around the sun. It rotates once in 10h - 13m - 59s. Its orbital velocity is 9.6 km/s (6.0 miles/s), its orbital inclination is 2° 29' 21", its orbital eccentricity is 0/056, its diameter at the equator is 120,535 km (74,914 miles), and its polar diameter is 108,728 km (67,575 miles). This difference is typical of all the outer planets because of their gaseous makeup and high revolution rate. Its density is 0.71 times that of water, its mass is 95.17 that of earth's, its volume is 1744 times that of earth's, its escape velocity is 32.26 km/s (20.05 miles/s), its gravity is 1.16 times that of earth's, and its surface temperature is -180° C. Saturn has many moons—forty-five at last count, recently recorded by the Cassini/Huygens probe in 2007. Of those, four are major moons, and the rest are much smaller. The major moons' data is (in order of size):

	Diameter	Distance from Saturn
Titan	5150 km	1,221,860 km
Rhea	1528 km	527,040 km
Lapetus	1436 km	3,561,300 km
Dione	1120 km	377,420 km

It is noted that their size has no relation to their orbital distance from Saturn.

One last bit of significance: While viewing the Cassini/Huygens special produced by NASA, images of the South Pole of Saturn showed an eye reminiscent of that of a hurricane on earth, directly over the South Pole. It had a round shape and was approximately three of earth's diameters in width. The Cassini space craft then went to the North Pole, and there was an opening similar to the one at the South Pole, but its shape was not circular. I could not believe my eyes; it was hexagonal, yes! A well defined six-sided figure, with a diameter of approximately five earths across. How could this be possible? A hexagonal shape cannot exist in nature; it is not natural, especially in this macro size. What could have caused this phenomena? The image causes my imagination to run wild (remembering the monoliths of Arthur Clarks, 2001). I know that NASA would not be using trick photography because that is unprofessional and unscientific. I'm eagerly looking forward to seeing more images taken from future probes sent to Saturn, especially of the North Pole, to hear the explanations given as to how this unnatural hexagonal shape can exist. One last point: I called J.P.L. Laboratories and spoke with an advisor, but he was not aware of the T.V. program that I watched that evening.

Titan

Satellite Data

Distance from Planet	1,221,860 km (759,228 miles)
Orbital Period	15.495 days
Orbital Velocity	
Orbital Inclination	0.33°
Orbital Excentricity	0.029
Diameter Equator	5150 km (3263 miles)
Density, Water = 1	
Escape Velocity	
Mass, Earth = 1	
Volume, Earth = 1	

TITAN

The second largest moon in our Solar System is Titan, at 3200 miles - 5100 kin. It is the largest of Saturn's moons. Titan possesses an atmosphere of nitrogen and methane, preventing sharp photographs of the surface from being taken. Titan was discovered by Christian Huygens in 1656. It has a rocky core and is surrounded by a mixture of water ice, ammonia, and methane. The atmosphere mentioned above starts at 60 miles with increasing density to the surface at a pressure of 1.5 times that of earth's (approximately 22 psi, although the surface is -168° C). It is thought to be in a liquid state, to a depth of over 1000 ft, as reported by Carl Sagan.

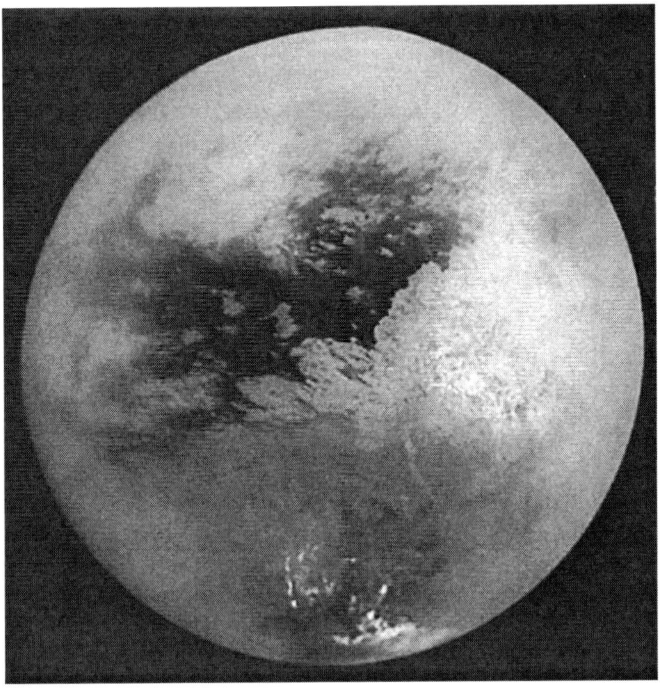

Robert E. Smolinski

Rhea

Satellite Data

Distance from Planet	527,040 km (327,487 miles)
Orbital Period	4.518 days
Orbital Velocity	
Orbital Inclination	0.35°
Orbital Excentricity	0.001
Diameter Equator	1528 km (949 miles)
Density, Water = 1	
Escape Velocity	
Mass, Earth = 1	
Volume, Earth = 1	

RHEA

As the second largest of Saturn's moons, and because it's located farther away than some of Saturn's major moons, Rhea shows far less surface change, of which is mainly pockmarked ice. It is the seventeenth moon from Saturn, and its location is just outside the famous rings of Saturn. Its heavily cratered surface suggests its age to be one of the oldest, and some of the craters are miles deep. Rhea's surface was recorded by Voyager 1 and shows great detail because of a lack of any atmosphere.

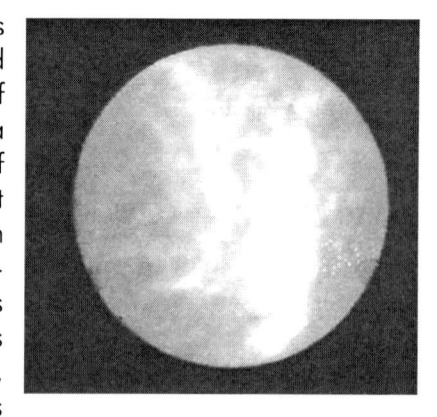

Dione

Satellite Data

Distance from Planet	377,420 km (234,517 miles)
Orbital Period	2.737 days
Orbital Velocity	
Orbital Inclination	0.02°
Orbital Excentricity	0.002
Diameter Equator	1120 km (696 miles)
Density, Water = 1	
Escape Velocity	
Mass, Earth = 1	
Volume, Earth = 1	

DIONE

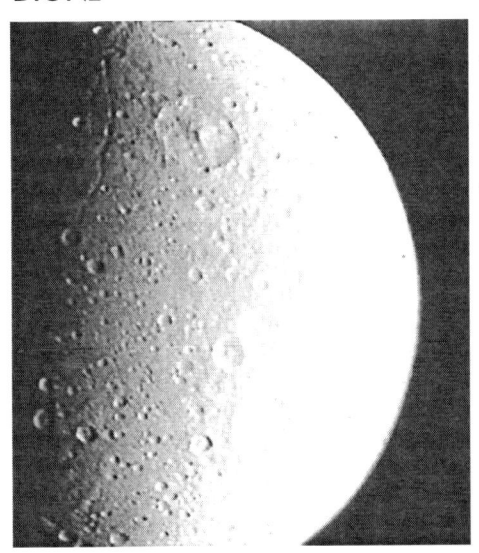

Dione is the second densest of Saturn's moons, probably because of its location closest to Saturn's rings; its surface has been recorded by Voyager 2 and shows extremely fine detail, also because of a total lack of atmosphere. It's composed of a mostly rocky and ice mix and shows numerous impact craters and fissures. Some of the craters are hundreds of miles in diameter and show evidence of being created many billions of years ago.

Eneceladus

Satellite Data

Distance from Planet	238,040 km (147,911 miles)
Orbital Period	1.370 days
Orbital Velocity	
Orbital Inclination	0.07
Orbital Excentricity	0.004
Diameter Equator	395 km (245 miles)
Density, Water = 1	
Escape Velocity	
Mass, Earth = 1	
Volume, Earth = 1	

ENCELADUS

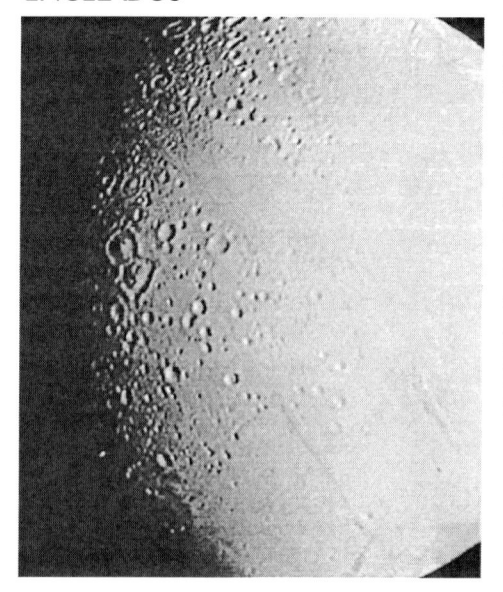

Number eight away from Saturn in orbit, recorded by Voyager 1, Enceladus has very sharp features that show areas of impact craters and also very smooth plains. It has an orange color and is thought to have subsurface water that occasionally wells up to obscure some surface features. It is almost twice the size of Mimas.

Mimas

Satellite Data

Distance from Planet	185,540 km (115,289 miles)
Orbital Period	0.942 days
Orbital Velocity	
Orbital Inclination	1.52°
Orbital Excentricity	0.020
Diameter Equator	190 km (118 miles)
Density, Water = 1	
Escape Velocity	
Mass, Earth = 1	
Volume, Earth = 1	

MIMAS

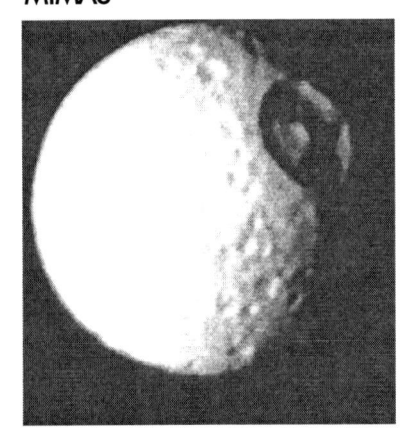

The seventh moon away from Saturn's surface in orbit was recorded by Voyager 1 and is unique because of the one large impact crater (80 miles in diameter). It is strange because the crater is about one third as large as the total diameter and has walls that rise three miles above the surface and is six miles deep, with a tremendous mountain about the same size six miles high and located dead center of the crater.

Tethys

Satellite Data

Distance from Planet	294,670 km (183,009 miles)
Orbital Period	1,888 days
Orbital Velocity	
Orbital Inclination	1.86 °
Orbital Excentricity	0.000
Diameter Equator	1046 km (650 miles)
Density, Water = 1	
Escape Velocity	
Mass, Earth = 1	
Volume, Earth = 1	

TETHYS

Tethys is the ninth in orbit from Saturn and is larger than moon number eight (Enceladus) by approximately 500 km. The surface composition appears to be all ice and has very few craters; it was recorded by Voyager 2, has one gigantic crater 250 miles in diameter, and a trench 1250 miles long running vertically north to south.

Lapetus

Satellite Data

Distance from Planet	3,561,3000 km (2,212,889 miles)
Orbital Period	79.331 days
Orbital Velocity	
Orbital Inclination	7.52
Orbital Excentricity	0.028
Diameter Equator	1436 km (892 miles)
Density, Water = 1	
Escape Velocity	
Mass, Earth = 1	
Volume, Earth = 1	

LAPETUS

The second-most distant moon of Saturn was recorded by Voyager 2, the surface of which has impact craters. This suggests a rock composition and is dark in places, and orange yellow in color. It is also the third largest of Saturn's moons.

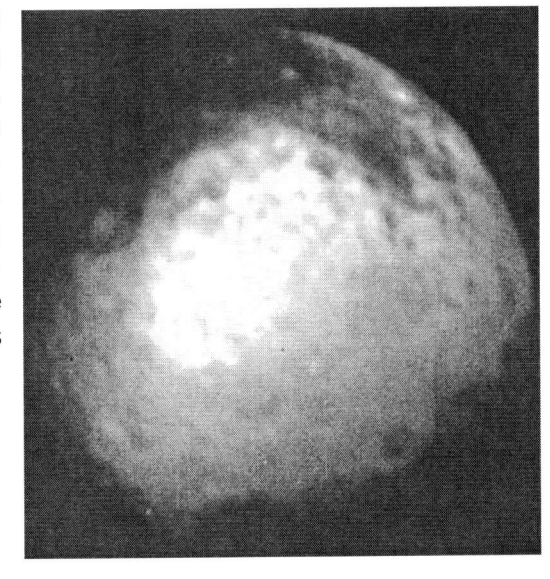

Robert E. Smolinski

Uranus

Planetary Data

Sidereal Period	30,684.9 days (84 earth years)
Rotation Period	17.2 hrs
Orbital Velocity	6.80 km/s (4.22 miles/s)
Orbital Inclination	0.733° / Axial Inclination 98°
Orbital Excentricity	0.047
Diameter Equator	51,118 km (31,770 miles)
Distance from Sun (mean)	2870 x 10^6 km (178,382,746 miles)
Density, Water = 1	1.27
Mass, Earth = 1	14.6
Volume, Earth = 1	67
Gravity, Earth = 1	1.17

URANUS

Planet seven, at 2,870,000,000 km (178,382,746 miles), was discovered by William Herschel in 1781, using his newly built telescope at Slough England. It's not possible to see the surface of Uranus, due to its dense atmosphere which is composed of water, ammonia, and methane, with a layer of helium above. Many writers refer to Uranus as a water world because of this property. Uranus takes 84 earth years to go around the sun once. It rotates on its axis in 17.2 hrs, its orbital velocity is 6.8 km/s (4.22 miles/s), its orbital inclination is 0.773°, its diameter is 51,118 km (31,770 miles). Its density is 1.27 times that of water, its mass is 14.6 times that of earth's, its volume is 67 times that of earth's, its gravity is 1.17 times that of earth's, and its surface temperature is -214° C (-353.2° F). If you know where to look, Uranus is just visible to the naked eye. Even with the aid of a telescope capable of high magnification, all that can be seen is a greenish disc. Of special interest is the fact that Uranus' axis has been altered, which is apparent when we see images taken by our space probes. Its axis is tilted 98°

toward the sun and could only have been caused by a collision or near collision with another stellar object. (In Chapter 4 I will describe in detail what I think caused this near catastrophe for Uranus.) Also a point of interest: While looking at the images of the moons of Uranus, taken by Voyager 2 in 1986, a thought occurred to me. It appears that at some time in the past, all twenty moons exhibited pockmarked impact craters, but only on one side. If this evidence were taken by itself just for one moon, it would not be of significance, but taken together with the 20 others, including the craters on our moon and Mercury, this suggests that perhaps two to three billion years ago there was a collision of such magnitude that it encompassed our total Solar System, envisioning massive violent interactions with other stellar bodies that could only be explained by galaxies colliding. I mentioned this in previous chapters, but I thought with this new evidence it was worth mentioning again.

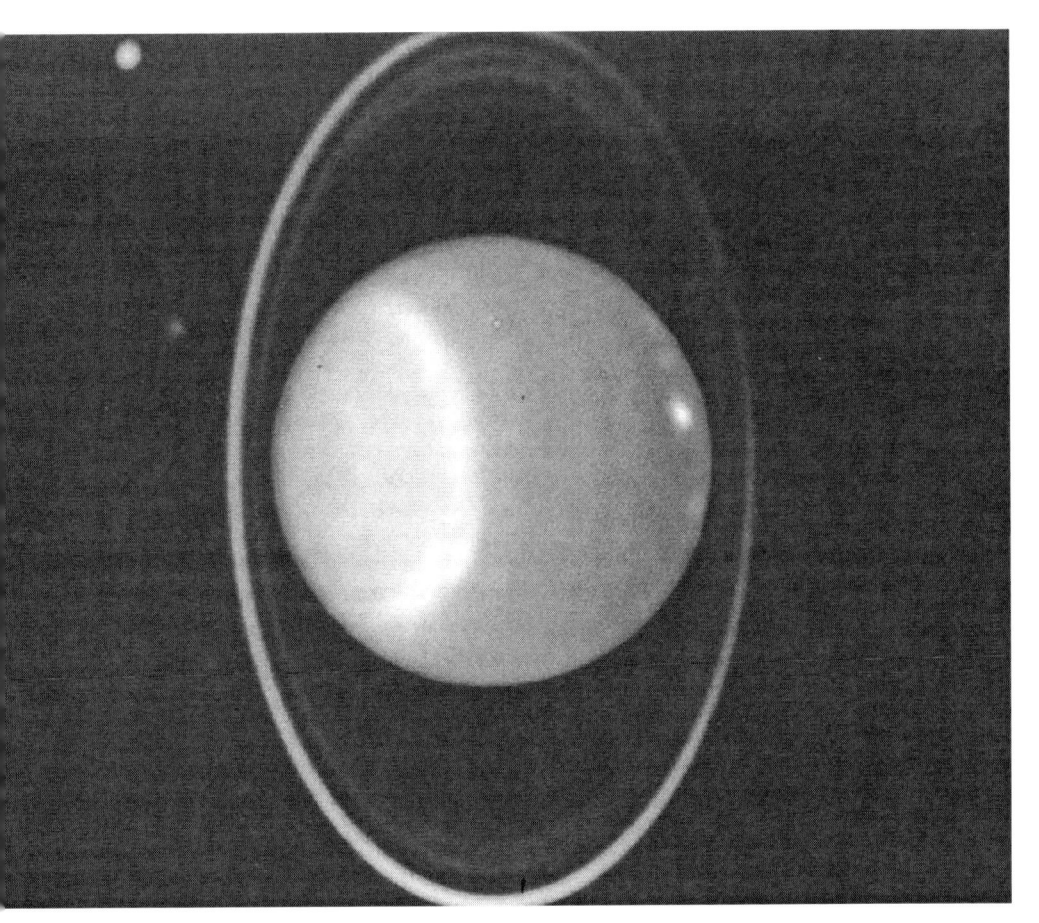

Because the moons of Uranus don't have an atmosphere (Titan being the only exception), they make great historical record keepers, unlike the planets because of erosion. Of the 20 moons of Uranus, four are major satellites. They are:

Diameter	Distance from Uranus
Miranda	1158 km
129,400 km	
Ariel	1169 km
191,000 km	
Umbriel	1578 km
256,300 km	
Titania	1523 km
435,000 km	

(More satellite data is available on individual moon data sheets.)

Voyager revealed rings that were also displaced by 98°, which were not visible from earth, and these rings dated approximately when the collision occurred. Five spots of considerable size were detected using radar, of which no other data is available. The bluish-green color of Uranus is caused by the absorption of red light and reflection of the shorter wavelength at the blue end of the spectrum. The methane in the upper atmosphere of Uranus is in the frozen state, and therefore creating a thick blanket, preventing the ammonia and water from escaping to space. Uranus also possesses a fairly strong magnetic field, which suggests a molten liquid core. There is also an unexplained difference between its axis tilt, 98°, and its magnetic axis, 58.6°, and is displaced by 4700 miles. Whatever was responsible for knocking Uranus off its axis is probably responsible for displacing its magnetic field also. There seems to be a similar anomaly on Neptune, which we visit next.

Miranda

Satellite Data

Distance from Planet	129,400 km (80,405 miles)
Orbital Period	1,414 days
Orbital Velocity	
Orbital Inclination	4.22°
Orbital Excentricity	0.0027
Diameter Equator	466 km (289 miles)
Density, Water = 1	1.3
Escape Velocity	
Mass, Earth = 1	
Volume, Earth = 1	

MIRANDA

The moon closest to Uranus was recorded by Voyager 2. Miranda was discovered by G.P. Kuiper in 1948. Its diameter, 500 km, placed it thirteenth in size of moons closer to Uranus. Its composition is thought to consist of a 55/45 mixture of rock and ice, which accounts for its density of 1.3. Its color appears to be gray, and its surface is unique because of two gigantic gouges at north latitudes, comprising one-fifth of the moon's surface. It was obviously caused by tremendous collisions with unknown objects.

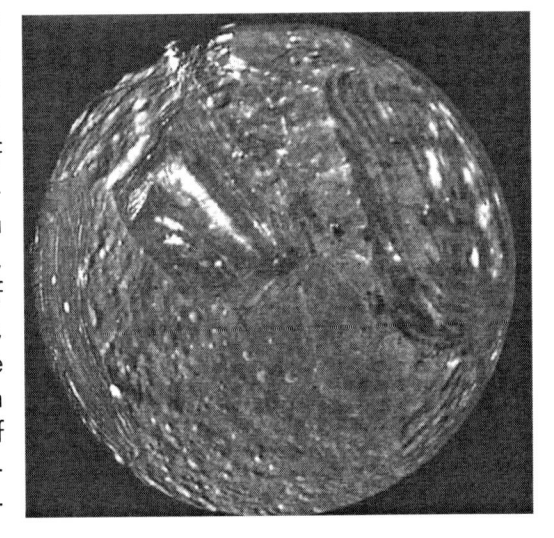

Ariel

Satellite Data

Distance from Planet	191,000 km (118,681 miles)
Orbital Period	2.520 days
Orbital Velocity	
Orbital Inclination	0.31°
Orbital Excentricity	0.0034
Diameter Equator	1158 km (719 miles)
Density, Water = 1	1.6
Escape Velocity	1.2 km/s
Mass, Earth = 1	
Volume, Earth = 1	

ARIEL

Ariel is located beyond Miranda, in orbit at about 190,000 km from Uranus. It was imaged by Voyager 2 and shows some impact craters but many faults, scarps, valleys, and surface cracks, suggesting tectonic activity. Its diameter is a little more than Miranda's, and its density is slightly greater at 1.6, which suggests a great percentage of rock to ice ratio.

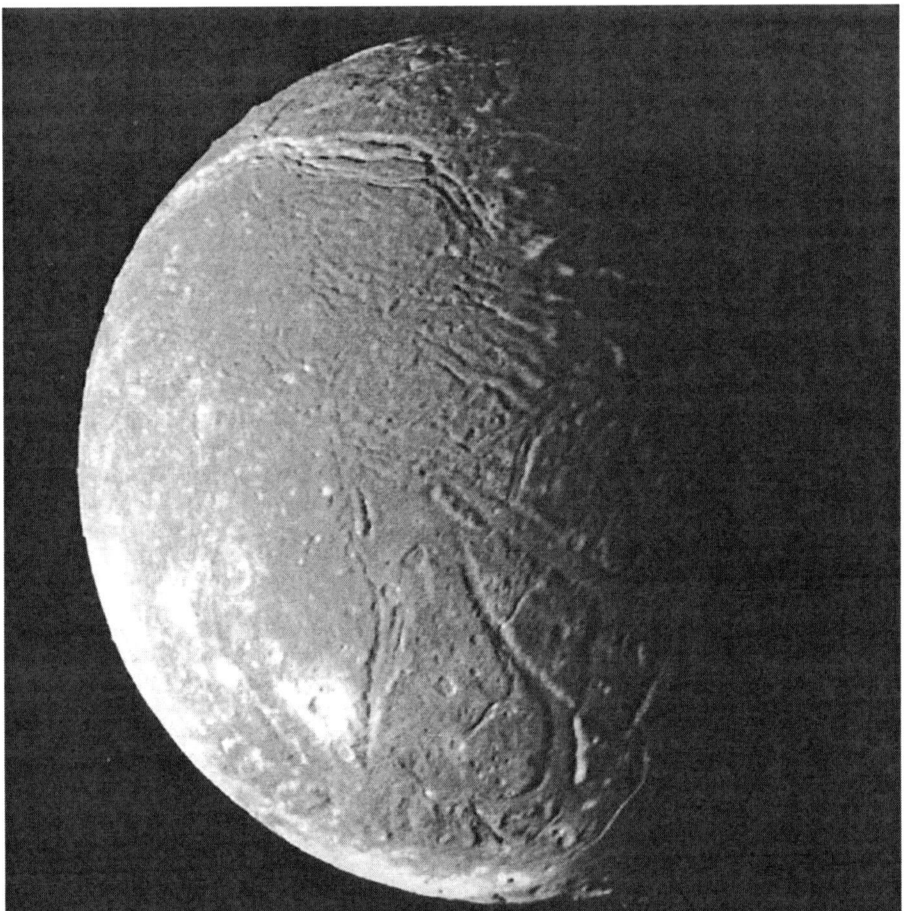

Oberon

Satellite Data

Distance from Planet	583,500 km (362,570 miles)
Orbital Period	13.463 days
Orbital Velocity	
Orbital Inclination	0.10°
Orbital Excentricity	0.0008
Diameter Equator	1523 km (946.3 miles)
Density, Water = 1	1.5
Escape Velocity	1.5 km/s
Mass, Earth = 1	
Volume, Earth = 1	

OBERON

Oberon is sixteenth in orbit from Uranus, was imaged by Voyager 2, and has a rocky surface. The composition of Oberon consists of a 50/50 mixture of rock and ice, which reflects its density of 1.5. Some of the images taken by Voyager 2 show high peaks on the order of three to four miles, although the number of moons of Uranus at last count is 20. Oberon is the last large moon; the next four range in size from 40 - 120 km. Oberon is heavily cratered, with many of the craters deep and dark, possibly caused by interior carbon.

Umbriel

Satellite Data

Distance from Planet	256,300 km (159,257 miles)
Orbital Period	4.144 days
Orbital Velocity	
Orbital Inclination	0.36°
Orbital Excentricity	0.0050
Diameter Equator	1169 km (726 miles)
Density, Water = 1	1.4
Escape Velocity	1.2 km/s
Mass, Earth = 1	
Volume, Earth = 1	

UMBRIEL

Umbriel is located approximately halfway between Ariel and Titania. Imaged by Voyager 1, its color appears dark green. Its surface is comparatively smooth and shows just a few impact craters, which are quite large, and a prominent frozen white equatorial cap. The largest crater, Skynd, is 68 miles in diameter, and the frozen equatorial region is 87 miles in diameter.

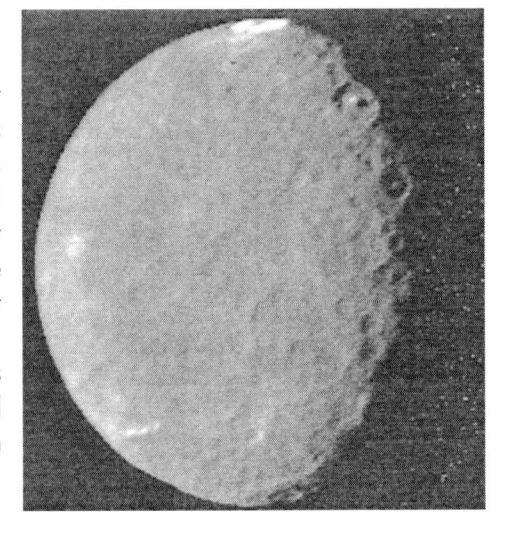

Titania

Satellite Data

Distance from Planet	435,000 km (270,296 miles)
Orbital Period	8.706 days
Orbital Velocity	
Orbital Inclination	0.014
Orbital Excentricity	0.0022
Diameter Equator	1578 km (980 miles)
Density, Water = 1	1.6
Escape Velocity	1.6 km/s
Mass, Earth = 1	
Volume, Earth = 1	

TITANIA

Titania is fifteenth in orbit around Uranus. Imaged by Voyager 2 in 1986, Titania shows a surface which consists of craters and ice cliffs. The color appears to be light brown, and the craters are white on a light brown background, suggesting that icy material below the surface was exposed on collision. Titania and Oberon share the privilege of being the largest moons of Uranus at approximately 1550 km, and because its density is 1.6, it must share the basic compositions of rock and ice that make up Ariel.

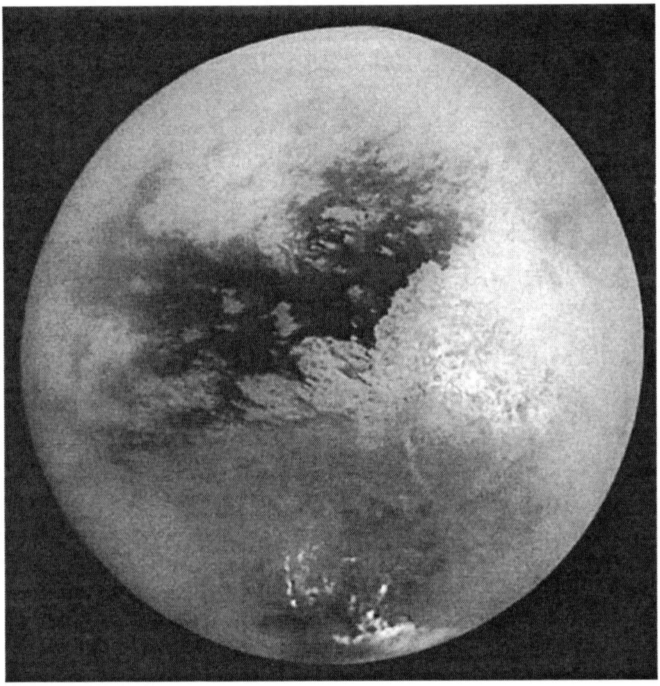

Neptune

Planetary Data

Sidereal Period	60,190.3 days – 164.8 Earth years
Rotation Period	16hrs-7m
Orbital Velocity	5.43 km/s (3.37 miles/s)
Orbital Inclination	1° 45' 19.8" / Axial Inclination 28.8°
Orbital Excentricity	0.009
Diameter Equator	50,538 km (31,402 miles)
Distance from Sun (mean)	4497 x 10^6 km (2,794,306,251 miles)
Density, Water = 1	1.77
Mass, Earth = 1	17.2
Volume, Earth = 1	57
Gravity, Earth = 1	1.2

NEPTUNE

Planet number eight, Neptune, is the last major planet in our Solar System, at a distance from the sun at 4,497,000,000 km (2,795,077,382 miles). Neptune was discovered by Johann Galle and Heindrich D'Arrest in 1846, after observing perturbations of Uranus' orbit. It definitely needs to be observed with the aid of a high magnification telescope and also reveals no surface details because of its upper atmosphere. The composition is similar to Uranus, primarily a mixture of hydrogen, helium, and methane, with the uppermost layer mainly methane in the frozen state. The best images we have as data were provided by Voyager 2 in its fly by in 1986 and Hubble's images in 1998. They show great detail, especially the ice clouds above the basic methane layer, marring the otherwise perfect blue atmosphere. It takes Neptune 164.8 years to circle our sun once. It rotates on its axis in 16 h - 7 m. Its orbital velocity is 5.43 km/s (3.37 miles/s), its orbital inclination is 1° 45' 19.8", and its diameter is 50,538 km (31,410 miles). Its density is 1.77 times that of water, its mass is 17.2 times that of earth's, its volume is 57 times that of earth's, it's escape velocity is 23.9 km/s (914.8 miles/s), and its surface gravity is 1.2 times that of earth's. Neptune has two major moons, Triton and Neried, and six lesser moons, all located within the major moon's orbit.

Diameter
Distance from Neptune
Triton   2705 km
354,000 km
Nereid   240 km
1,345,000 km - 9,688,500km

Neptune's moons, eight in all, are arranged in order of size. The smallest is the nearest, and the largest is the farthest. Images taken by Voyager 2 show surface details of the moon so fine that latitude and longitude maps were generated to measure and name many of the surface features, especially Triton, which exhibits the most contrasting colors from deep blue in the northernmost parts to light pink in the southern polar regions, with areas of orange in between. Voyager was able to measure the surface temperature of

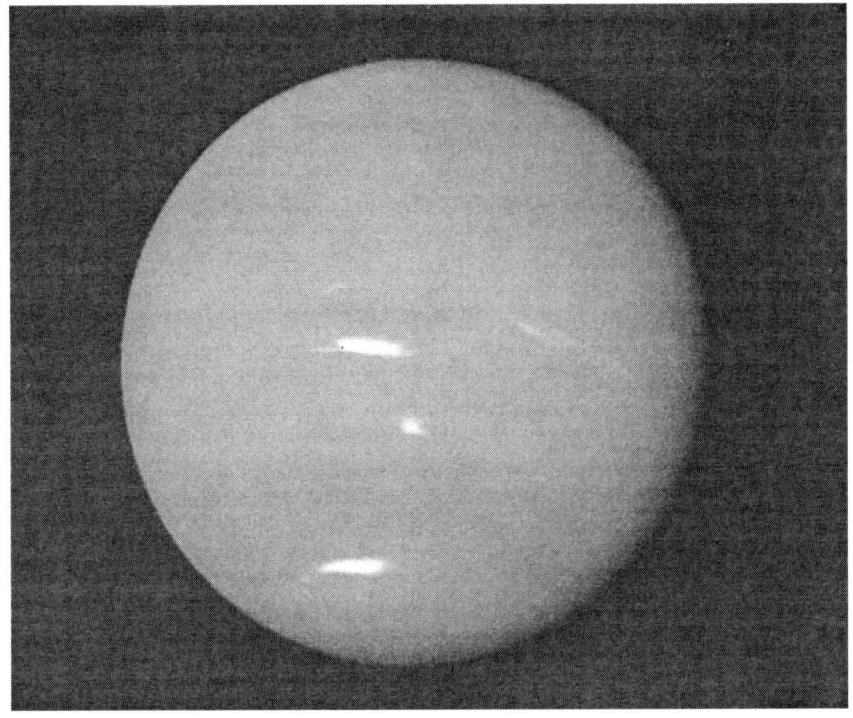

Triton, which is 236° C (-392.8° F), the coldest of all bodies in our Solar System ever recorded. Triton also exhibits what are believed to be nitrogen geysers caused by liquid nitrogen being forced up to the surface from pockets one hundred feet below. The velocity of the geysers was measured at 500 ft/s, allowing the nitrogen jets to overcome the escape velocity of 0.9 miles/s, sending Geyers up many kilometers before being attracted by Triton's gravity and falling back to the surface. These Geyers carry darkened material from below, after venting, and deposit that material downwind from the vent, as was recorded in the images near the south pole of Triton. Also, to describe the bluish mid-region, the so-called cantaloupe terrain (it looks like the skin of a cantaloupe), it consists of a multitude of random geometric patterns, intersecting at approximately 20 miles across. Lastly, Triton's northern hemisphere is unique in appearance because of rugged circular depressions, resembling flattened impact craters with varying diameters— between 1 and 30 miles across—which have never been seen on any other surface in our solar system. Triton's orbit is nearly circular (0.0002) and Nereid's highly eccentric (0.749). Triton's orbital inclination is equal to 159.9°, and Nereid's is 27.2°. If you can imagine Neptune's orbital plane being 180°, then Triton's orbit approaches that of Neptune's, as opposed to that of Nereid's orbit, being tilted in the opposite direction such that it approaches Triton's polar axis. These exaggerated physical parameters do not obey the laws of conservation of energy and momentum and obviously were caused by some force of disruption, once again suggesting a collision or near collision at some time in the past.

UNDERSTANDING THE UNIVERSE

TRITON

Satellite Data

Distance from Planet	354,800 km (220,462 miles)
Orbital Period	5.877 days
Orbital Velocity	
Orbital Inclination	159.9° (what caused this)
Orbital Excentricity	0.0002
Diameter Equator	2705 km (1681 miles)
Density, Water = 1	2.0
Escape Velocity	
Mass, Earth = 1	
Volume, Earth = 1	

TRITON

The seventh of eight moons in orbit from Neptune, and by far the largest, at 2705 km in diameter, is Triton. It was imaged by Voyager 2 and exhibits the most interesting surface features of any body in our Solar System, the compositions and colors of which are not duplicated on any other body. From the blue smooth grapefruit-like surface of the northern three quarters to the white mountainous cratered and fissured south quarter of its surface, and also because of its retrograde motion opposite that of the rotation of Neptune, Triton is unique again. Its density is 2.0, which means its compositions is more rock than ice, and it has the coolest temperature recorded for any body in our Solar System: -236°C.

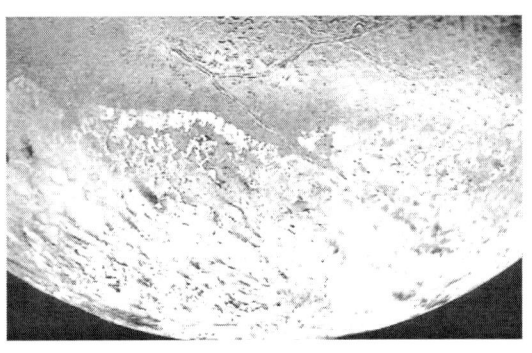

Pluto

Planetary Data

Sidereal Period	90,465 days (248.4 Earth Years)
Rotation Period	6d-9h-17m
Orbital Velocity	4.7 km/s (2.9 miles/s)
Orbital Inclination	17.2° / Axial Inclination 122.5°
Orbital Excentricity	0.248
Diameter Equator	2324 km (1444 miles)
Distance from Sun (mean)	5900 x 10^6 km (3,666,090,034 miles)
Density, Water = 1	2.02
Mass, Earth = 1	0.0022
Volume, Earth = 1	0.01

PLUTO

For the last 77 years, Pluto has been considered the ninth planet of our Solar System. In 2007, the astronomical community decided to downgrade Pluto to a planetesimal. I understand their reasoning to a degree because some of the parameters necessary to define a planet include not clearing debris from its orbit or its orbit being too eccentric, Nevertheless, I elect to include Pluto as a planet in our system because of all the hard work and many years devoted to the cause by Percival Lowell and Clyde Tombough, two great astronomers of note. Some of the negative reasons for electing to downgrade Pluto include being too far away, too small, and having an orbit too elliptical compared with the rest of the planets. I personally don't think these reasons have merit. At this point I would like to describe to the reader the dedication to the field of astronomy of Lowell and Tombough. In the 1920's Lowell tried to find the cause of Neptune's eccentric orbit, but to no avail. Then in 1930 Tombough discovered the object responsible, only after spending many years of taking and examining many photographic plates and then presenting those plates as evidence to the astronomical community. Finally, after gaining acceptance of his discovery, Tombough proceeded to name the object in question Pluto, after the god of the underworld (described in Dante's *Inferno*) because of the similarity it shared and the gloomy place it occupied in space.

Many theories about Pluto abound, one of which was that a time long ago, Pluto and Triton were satellites of Neptune, but for some unexplained reason Triton stayed captured and Pluto was not. This is a possibility, but without some evidence, or at least a hypothesis, it can only be considered an idea or fable (paralleling the flat earth theory of old).

Pluto's mean distance from our sun is 5,900,000,000 km (3,666,090,034 miles), and because of its highly eccentric orbit, I will state its maximum and minimum distance because of the controversy associated with this parameter. Its maximum distance is 7,375,000,000 km (4,582,612,543 miles), and its minimum distance is 4,425,000,000 km (2,749,567,526 miles) for an orbital eccentricity of 0.6%. It takes Pluto 247.57 years to orbit the sun once. It rotates on its axis in 6 d – 9 h – 17 m. Its orbital velocity is 4.7 km/s (2.9 miles/s), its orbital inclination is 17.2°, its mass is 0.0022 that of earth's, its escape velocity is 1.18 km/s (0.7 miles/s), its surface temperature is -230° C (-382° F), and its diameter is 2334 km (1444 miles).

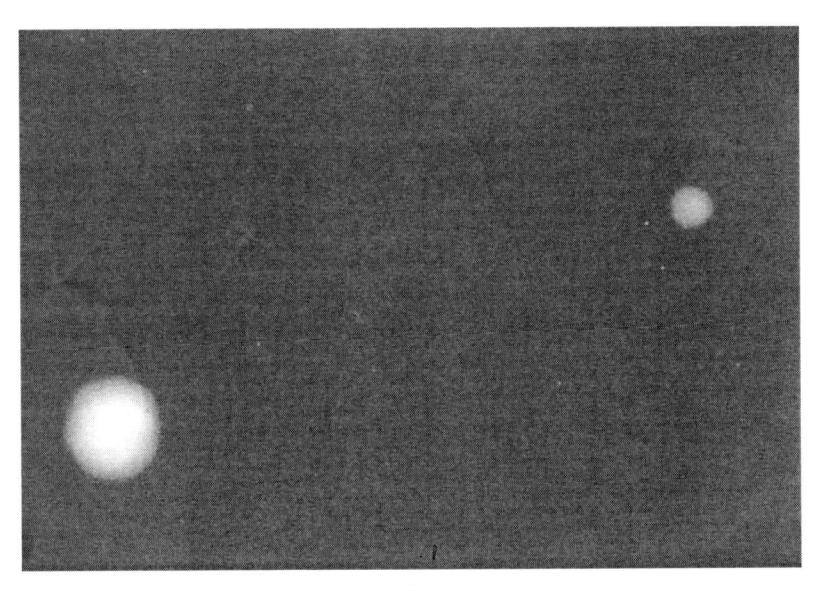

In 1977 Pluto was found to have a companion, Charron, named after the Boatsman of Donnate's fame, moving together, although separated by 12,500 miles. Charron was verified by images taken by Hubble in 1994, which revealed its diameter to be only 790 miles with a mass one-twelfth that of Pluto's. Using spectroanalysis, the atmospheres of Pluto and Charron were found to consist of methane and nitrogen, with Charron showing signs of water ice.

The Kieper Belt

Beyond the orbit of Pluto there exists a vast expanse which contains many bodies of planetesimal size and smaller. Many of their orbits are highly eccentric and in certain instances come as far in as the orbits of Saturn, such as 5145 Pholus, which was discovered by D.L. Robinowitz in 1992, when he was at the Kitt Peak Observatory in Arizona. Little is known about the Kieper Belt, primarily because of its extreme distance, typically 5.5×10^9 miles, placing visual limitations on our telescopes and instruments just to resolve its components. Another of the objects discovered by David Jewitts and Jane Luu in 1992 at the Mauna Kea Observatory in Hawaii was QB1, revealing an eccentric orbit of 3170×10^{46} miles by 4100×10^6 miles. Its period of rotation was estimated at 296 years, with a diameter of less than 100 miles.

Some comets are also thought to originate in the Kieper Belt, but the majority are believed to come from an area still farther out called the Ort Cloud.

Ort Cloud

The Ort Cloud is thought of as being the birthplace of most, if not all, comets. The majority of the icy bodies occupying the area are at a mean distance of one light year. This is approximately 5.88×10^{12} miles. Everything known about this area is theoretical in nature because of the vast distance. Regarding the size of the objects in question, a commonly accepted scenario is considering a particular comet's actions—for example, it might have a highly eccentric orbit swinging around the sun, with a period of millions of years, and could continue in its orbit, be pulled into the sun, or sucked into Jupiter, as the Shoemaker Levy Comet was just recently, in 1994. One of the closest comets that we are tracking is Comet Pizarro, now located in the Asteroid Belt. It has been identified by its pronounced tail. When we consider the Ort Cloud as where most comets originate, I believe we should not discount the possibility that this is the place were a giant comet made mostly of ice originated some three billion years ago, and this comet with an elliptical orbit around our sun eventfully ended up colliding with earth. This explains one of the questions no one seems able to answer: Where did all the water in earth's oceans come from? This scenario as stated makes the most sense to me because it is the only theory that could possibly explain the presence of the mind-boggling amount of water that covers the earth's surface (which, by the way, covered more like 90% of the earth's surface, and today the best estimate is 78%). So another interesting bit of information is disclosed—if, three billion years ago, the earth's surface was 90% covered, and today it is 78% covered, then in three billion years earth lost 12% of its surface water. Projecting forward, we can then estimate at least another three billion years before we lose another 12%, so 78% today minus 12% equals 66%. That leaves us with 66% surface water, and if 78% is equal to 10,000 quadrillion gallons, then that leaves us with 8800 quadrillion gallons, which should leave us with enough water until the sun turns into a Red Giant, at which point I don't think water will matter at all.

CHAPTER 2 - EARTH'S PLACE IN OUR SOLAR SYSTEM

Many articles have been written about the earth and the possibility that there are other earths like ours out there in space. I think not! Let me present my argument. How many times have you heard the argument, "Just look up at night; the stars you see with your naked eye are only a small percent of the total stars in the universe," and they go on to say that with the aid of a telescope, you could multiply what you see by 10^100. They continue that if only 1% of the stars supported planets, and those had all the conditions to support life (this is the kicker—had all the proper conditions to support life), there could possibly be thousands of planets like earth out there. I say to those who think this sounds convincing: It appears you have not done your homework. Consider the conditions that have to exist in order for life as we know it to flourish: gravity, water, temperature, atmosphere, sun type, orbital stability and distance, mineral content, radiation gradient, seasons, and earth's interaction with other celestial objects. Because all of these conditions are required to sustain life as we know it and are taken for granted, and because most of the population of earth does not possess the scientific knowledge to understand how these conditions affect us, I will attempt to define these conditions more completely.

1.) Gravity—First we must use earth's gravity as a reference. Therefore we set (G=1), the reference gravity. It is responsible for all life on earth, from the molecular to the macro level. To get a feeling for how gravity affects life on earth, I will endeavor to use the argument of limits (the maximum and minimum). Let's consider maximum surface gravity on Jupiter. Jupiter's gravity is so great because of its mass. Hypothetically, if we were to be transported to the surface of Jupiter and all other variables notwithstanding, we would be crushed by the G force of 2.64 times that of Earth. A 200-pound man would weigh 528 pounds. We would not be able to move; breathing would be extremely difficult, and our circulation system would not be able to supply blood to our organs. We would die. Now let's look at the other extreme, the minimum gravity found on our space station in earth's orbit. The gravity is equal to "0" (weightless). Just a few of the problems found were calcium leaking out of our bones, muscle atrophy, disruption of our balance system (semi-circular canals in our ears), not to mention normal eating, the passing of waste, and the fact that "0" gravity cannot support an atmosphere to breathe, making it necessary for a closed system capable of supplying oxygen at the necessary pressure. These are just the bare essentials that gravity would affect. I'm sure the reader could come up with more if he or she tried.

1.) Water—Without water I think we all agree that life would be impossible. Even today, with all the knowledge of our greatest minds, no one has put forth a theory of where our water came from; the best we can do is the Comet Theory.

2.) Temperature—A sustained temperature of greater than 212° F or less than 32° F is also fairly obvious. Primarily, if a temperature of 212° F is sustained, we would boil; if a temperature of less than 32° F is sustained, all water would freeze, liquid water would not exist, and neither would we.

3.) Atmosphere—Our atmosphere on earth is composed of 78% nitrogen, 21% oxygen, and 1% trace gases. The O_2 component is absolutely essential, not only as of a percentage but also as a density, and must be maintained at 14.7 psi at sea level, or our bodies would cease to function (hypoxia).

4.) Sun type—The sun must be a G2 type sun (see Chapter 6 for an explanation of star types), or life as we know it would not be possible. Why? Because the temperature and physical size must be exactly equal to our sun in order to maintain not only the temperature, but the spectrum of light necessary to support all life (plants, animals, and us).

5.) Orbital eccentricity—Another of the important variables of earth is its orbital eccentricity. In its path around the sun, earth scribes an ellipse, not a perfect circle. Its perigee (closet point) is 147×10^6 km (91,361,109 miles), and its apogee (farthest point) is 152×10^6 km (94,468,614 miles), giving it an orbital eccentricity of 0.017. Surprisingly, the perigee occurs in the northern hemisphere in winter and its apogee in our summer. To justify how important it is that this variable be maintained, consider the argument once again of the limits—too close we would cook, and too far we would freeze.

6.) Mineral content—The mineral content of earth is such that almost everything is made up of some combinations of 96 natural elements, with millions of compounds. To date, no one has come up with a formula to explain how these basic elements combine to explain the basic building blocks of life—the DNA molecule.

7.) Radiation gradient—Consider the sun as a source of life and light, without which there would be none. This light consists of a component of radiation associated with it once again—too much we fry, not enough we die. The amount of solar radiation has to be just right.

8.) Earth's axis—Earth's axis has a tilt of 23.5°, which is responsible for our seasons. If there were no tilt, then the poles would always be frozen, and the equator would be greater than 150° F. There would only be a small band between the poles and equators at approximately 45° (N+S) latitude that could support life.

9.) Earths interaction with other objects in our Solar System— How fortunate we are to enjoy the gravitational protection of our sun and the other planets to pull in objects flying through our Solar System and preventing them from colliding with earth, some of which have done so in the past and will likely do so in the future, but hopefully not in our lifetime. We recognize what happened 65 million years ago to creatures much larger and stronger (the dinosaurs) because of just a single impact that created the Gulf of Mexico and laid waste to 98% of all life on planet earth. But thank God the mammals survived! So taking any one of the ten basic variables formulated to their limits, as I have described, could cause life to be extinguished. Now, combing these ten variables together, you can get some perception of the odds necessary to keep all these variables within some realm of consistency, and you can now appreciate how unique our planet earth really is. I hope by painting this mental picture, you will agree with me that the earth is a very special place, and whose benevolence is responsible for this convenient protection I will leave for another book. But I will say this: There is no way anyone can make a blanket statement about there being millions of other planets "just like earth" out there in space, and by the way, the premise that there are other planets out there suggests that in an emergency we could go there. Let me ask a simple question. Assuming you could find a planet out there, how would you get there? This presents a problem in itself (see Chapter 12, in which I discuss the problems of space travel).

CHAPTER 3 - THE HISTORY OF OUR ASTRONOMERS, SCIENTISTS, AND MATHEMATICIANS

I think it's necessary for the reader to get an idea of how astronomy and science advanced and how this brought us to the place we are now. I will endeavor to relate in general names, dates, and accomplishments of the most noted astronomers, scientists, physicists, and mathematicians who have laid the groundwork that helped us better understand our earth, Solar System, the universe, and the physical laws that control their actions. I have selectively chosen to present these individuals in three groups: the ancients, the astrologers, and the enlightened.

The ancients existed in the age before Christ (B.C.). Due to the destruction and burning of the library at Alexandria and all written records that were kept there, word of mouth and the contributions of those that were made before this horrendous event were all we had to go on. So with this fact in mind, let's proceed. For the reader's convenience, I would like to point out that the classifications of astronomy did not exist before 570 A.D., when Isidorus distinguished between astrologers and astronomers. Prior to those times, the ancient Greeks were referred to as Philosophers. So let's get started.

Thales of Miletus (636 - 546 B.C.) was a Greek who was regarded as the first western Philosopher. His most profound contribution was a belief that all things were composed of what he referred to as the universal substance—water.

Anaximander (611 - 547 B.C.) was a Greek who, unlike Thales (his teacher), thought the universal substance was boundless or indefinite. He did not believe a single substance underlies all things.

Anarigoras (500 - 428 B.C.) was the teacher of Socrates who proposed an infinite number of unique particles of which all things were made.

Anaximenes (sixth century B.C.) believed all things are made up of a single universal substance—air.

Empedocles (495 - 435 B.C.) was a Greek who believed the universal substance consisted of four elements—air, fire, water, and earth. He thought love and hate caused the mixing of the elements.

Pythagoras (582 - 507 B.C.) was a Greek who believed numbers constituted the true nature of all things.

Parmenides (549 - 484 B.C.) was a Greek who believed "being" was the basic substance of which all things are composed of, and that motion, change, time, difference, and reality were all illusions of the senses.

Socrates (464 - 399 B.C.) was a Greek and the teacher of Plato. He had no specific ideas about a universal substance, but he steadfastly believed the unexamined life is not worth living. He was brought up on charges of corruption and heresy, sentenced to death, and committed suicide by drinking poison.

Plato (428 - 348 B.C.) was a Greek and a student of Socrates. He authored *The Republic*, where philosophers and kings trained at the highest level of moral and mathematical knowledge. His views on cosmology influenced the next 2000 years of scientific thinking.

Aristarchus of Miletus (190-270 B.C.) was a Greek and the first to propose the heliocentric theory of our Solar System.

At this point I would like to mention that the previous list of contributions to our subject are all Greeks. Although some of the ideas were mystical in nature, others were profound and deserve our recognition as being some of the greatest minds and teachers ever. We must also take into consideration that the Scientific Method had not yet been invented. Because of this, kudos must be given to some for original thought. Next I would like to describe the following group—astrologers, a mystical group, not of good science, that surprisingly dominated the period from 120 A.D. to 1400 A.D., the period referred to by many as the Dark Ages, primarily because nothing of scientific value had been gained over that period. It started with the first Ptolemy, of which there were many culminating in A.D. 140. They generated star charts for use in navigation of the trading ships of that time. These charts were useful, allowing sailors to find their way from country to county by the shortest distance over oceans and seas. Although the charts were practical they were not accurate, and at times ships would have to sail 200 miles or more before finding the ports they were looking for (by following the shore line). The star charts of the Ptolomy era, although contributing much by virtue of the book *The Almagest*, in which many constellations and stars were plotted and named, also contained one of the biggest misconceptions since the advent of the Flat Earth Theory—that of "Epicycles," Ptolemy's misguided attempts to explain retrograde motion of the planets. Surprisingly, this was believed for the next 1500 years.

In A.D. 1270 King Alfonso, the tenth of Castile, spent ten years translating *The Almagest*, to what is now referred to as the Alphonsine Tables, but also included the false "Epicycles" to explain the still not understood retrograde motions of the planets.

The time from A.D. 1000 - 1400 was lost to any understanding of why things were as they were. But by A.D. 1500 knowledge and understanding started to show their heads. Improvements started to come forth. Christopher Columbus didn't fall off the end of the earth, and Magellan returned home from his voyage. And with the invention of the Gutenberg Press, works of the Greeks and other scientists became available to many, and new ideas and theories began to come forth. This was the start of what I refer to as the third period of Enlightenment—or, the more popular name, the Renaissance. In 1543 Copernicus, on his deathbed, released his book *De Revolvations Orbitum*, describing the heliocentric Solar System that Aristarchus envisioned 1800 years earlier. Copernicus included the "Epicycles" of Ptolemy's fame in his book because he still did not understand retrograde motion, and it wasn't till Johannes Kepler, influenced by the teachings of Copernicus that finally made sense, of the true nature of retrograde motion of the planets, putting "Epicycles" to rest.

In 1546 – 1601 Tyco Brahe, using mechanical sighting devices, generated extremely accurate star charts because telescopes had not yet been invented. His venture was financed by the Queen of England, primarily because the English sailors were having the same trouble navigating because of a lack of accuracy in the star charts. Tyco charts were so accurate they had resolutions of less than a minute of arc—an outstanding feat considering his equipment. His work was done at his observatory at Hven, 20 miles northeast of Copenhagen Demark.

Next in order is Johannes Kepler (1571 - 1630). Using the groundwork laid by Copernicus, Kepler made many discoveries of significance of which elliptical planetary orbits, their descriptions, formulas, and mathematics are still used today to describe how planets move, such as Kepler's Law of Planetary Motion, and, finally, describing the true reason for retrograde motion of the planets due to "Epicycles." After his explanation, which was the angular difference with respect to time of viewing and was responsible for the apparent change in the positions of the planets being viewed, the term "Epicycles" was never again used to describe retrograde planetary motion.

Isaac Newton (1693 – 1727) was probably the greatest astronomer in history, primarily because of his unique contributions, inventions, and discoveries, some of which are the reflection telescope, the spectrum of light, the fluxion (calculus), gravity theory, equations theory, laws of motion, and so many others that would require a book on its own to cover all the bases.

At this point I would like to present a list of our ancestors in science, their contributions, and a short description of their accomplishments and discoveries.

Date	Culture / Person	Accomplishments / Discoveries
100,000 B.C.	Caveman	Cave drawings
3000 B.C.	Egyptians	Pyramids
2500 B.C.	Chinese	Celestial records
624 B.C.	Thales (Greek)	Planetary motion
270 B.C.	Aristarchus (Greek)	Heliocentric Solar System
260 B.C.	Erathostenes (Greek)	Round earth - measure diameter
A.D. 120	Ptolemy (Greek)	Almagest - Epicycles (incorrect)
A.D. 510	Isidorus (Greek)	First to coin the word astronomy
A.D. 813	Arabic	Beginning of Dark Ages
A.D. 1433	Ulugh Beign	Observatory at Sanarkand
A.D. 1449	Astronomy	Collapse
1473 – 1543	Copernicus (Polish)	Resurrection of astronomy
1564 – 1642	Galileo (Italian)	First practical telescope
1571 – 1630	Kepler (German)	Planetary motion-retrograde motion
1629 – 1695	Huygens	Designed telescope-discovered Uranus
1643 – 1727	Newton	Reflection telescope-Law of Motion, etc
1736 – 1813	Lagrange	Mathematical formulas-"Lagrange Point"
1821 – 1894	Helmholtz	Conservation of energy
1824 – 1887	Kirchoff	Spectrum analysis - optics - electricity
1826 – 1866	Riemman	Geometry - mathematics - tensor
1831 – 1879	Maxwell	Electro - Magnetic Theory - Gas Theory
1644 – 1710	Roemer	Light of speed
1857 – 1894	Hertz	Radio waves
1858 – 1947	Plank	Complex numbers - tensor - constant
1857 – 1894	Hertz	Electric theory
1879 – 1955	Einstein	Photo electric effect - $E = MC^2$
1885 – 1962	Bohr	Atomic structure - radiation
1887 – 1961	Schodinger	Atomic theory
1901 – 1976	Heisenberg	Quantum mechanics-uncertainty principle

UNDERSTANDING THE UNIVERSE

I would also like to include a list of scientists who laid the groundwork for the fame that many of the above scientists were given most of the credit for, not to mention the Nobel prize, without which I don't believe would have made the discoveries they were famous for, in alphabetical order.

Date	Name	Accomplishments / Discoveries
75 – 1836	Ampere	Electric current laws
76 – 1856	Avogadro	Avogadro's number: 6.023 x 10^23
00 – 1782	Bernouli	Gas laws
27 – 1691	Boyle	Gas formulas
46 – 1601	Brahe	Star charts (no telescope)
81 – 1868	Brewester	Refraction
31 – 1810	Cavendish	Hydrogen and gravity measurements
04 – 1990	Cherenkov	Radioactive blue glow in water
67 – 1934	Curie	Radium
96 – 1650	De Cartes	Mathematics, analytical geometry
02 – 1984	Dirac	Magnetic (monopole)
0 B.C.	Euclid	Logical geometric axions
07 – 1783	Euler	Mathematics - fluid flow
91 – 1867	Faraday	Electric theory - induction
01 – 1665	Fermat	Refective-math (theorm $A^n + B^n = n<2$)
00 – 1860	Fizeau	Light speed - rotating strobe
60 – 1920	Foucault	Pendulum-rotating mirrors, light measurement
68 – 1830	Fourier	Transform mathematics
06 – 1790	Franklin	Lighting-current flow-printing-spectacles
89 – 1870	Farnhofer	Lens - diffraction grating
88 – 1827	Fresnel	Lens - Wave Light Theory
64 – 1642	Galileo	Telescope
77 – 1855	Gauss	Electron charge-magnetic flux-mathematics
44 – 1603	Gilbert	Earth magnetic poles
82 – 1945	Goddard	First liquid fueled rockets
50 – 1925	Heavyside	Electromagnetism
01 – 1976	Heisenberg	Uncertainty principle
21 – 1894	Helmholtz	Conservation of energy
97 – 1878	Henery	Laws of induction
57 – 1894	Hertz	Oscillators - CPS - radio
73 – 1967	Hertzsprung	Diagram
24 – 1924	Hewish	Pulsars
35 – 1703	Hooke	Spring oscillators
29 – 1695	Huygens	Reflection telescope -Uranus–Saturn's rings

1822 – 1907	Kelvin	Thermodynamics - scale
1824 – 1887	Kirchoff	Electric laws - Norton's equivalents
1736 – 1813	LaGrange	Partial differentials - fluid flow
1749 – 1827	La Place	Electric laws-math-transform-spectrum
1743 – 1794	Lavoisier	Chemist - executed in reign of terror
1646 – 1716	Lentz	Induction - EMF
1853 – 1928	Lorentz	Radiation - magnetism
1836 – 1916	Mach	Sound 1100 ft/s
1831 – 1879	Maxwell	Electromagnetic Theory - equations
1852 – 1931	Michelson	The Either
1838 – 1923	Morely	The Either
1642 – 1727	Newton	Fluxion-ref telescope / spectrum / gravity
1789 – 1854	Ohm	Resistor Laws - Ohm's Law
1858 – 1947	Plank	Electron charge 6.6252×10^{-27} nm
323 B.C.	Ptolemy	Almagest - false Epicycles
1826 – 1866	Reimann	Tensor-mathematics complex numbers, et
1842 – 1919	Rayleigh	Mathematics - gas density - argon gc
1900 – 1985	Richter	Pendulum - gravity
1675 – 1742	Roemer	Light speed
1871 – 1937	Rutherford	X-Ray
1887 – 1961	Schrodinger	Atomic Theory
1591 – 1626	Snell	Reflective telescope
1856 – 1943	Tesla	Wireless power transmission
624 – 546 B.C.	Thales	Geometry - pre-astronomy
1773 – 1829	Young	Wave Theory-3 color, light Rosetta Stone

I would like to include some of the more recent contributions to science and astronomy, in no particular order.

Name	Accomplishment / Discoveries
Carl Sagan	Author - astronomer - theorist
Steven Hawkings	Astronomer royal - author - theorist
Fred Hoyle	Astronomer - author - theorist

CHAPTER 4 - UNDERSTANDING EARTH THROUGH TIME

"Using Our Imagination"
Let's take a ride in our time machine. Set the clock to 4.7 billion years ago. As observers we have nothing to fear. The destination is our Solar System, and what do we see? Our earth condensing from the accretion disc of matter in our Solar System. Initially earth was a hot molten rocky blob of material, so we set our controls to move ahead. After 500 million years the earth has cooled to a point, where a solid crust has formed at the surface and as time passes, and the cooling continues, we now see a well defined crust, a solid surface, a somewhat hotter mantel and a liquid molten core, which we were able to detect using our special onboard instruments. We move ahead to a point when earth was approximately one billion years old. Our scanner detects a tremendous ice comet that originated in the Ort Cloud, heading directly on a collision course with earth. There was an impact of cataclysmic proportions, and after the dust settled the earth was the recipient of 386 quadrillion metric tonnes of H2O. (We had water.)

    It took another 100 million years for life to develop in the primordial soup, and eventually biological forms of life started to flourish. As the complexity of life forms increased, higher orders of life appeared, such as Theotolites, producing oxygen by photosynthesis and creating an atmosphere rich in oxygen. Over the next three billion years, millions of complex marine life flourished, and after another 500 million or so years, oxygen levels were rich enough where plant life began to appear on land. Eventually the sea creatures developed limbs from fins and started to walk on land (after transforming gills to lungs), then after another 100 million years, reptiles appeared and ruled the early Paleozoic period. At this point there was another collision, only this time the mass of the visitor was such that it shook the earth to its very core. Within a very short time earth lost 90% of all living creatures. Finally, after a recovery period of 50 million years, the first dinosaurs appeared and dominated the planet for the next 250 million years. Then, about 65 million years ago, there was another impact in the Gulf of Mexico. The impact was of such force that it threw surface material out into what is now the Atlantic Ocean, creating the islands of Cuba, Haiti, Dominican Republic, and all the smaller islands in the area we see today. That material thrown into the upper atmosphere and out to space created what is now referred to as a nuclear winter. This disaster caused 80% of all plant life to die, which led to the death of 100% of all the dinosaurs. The surprising thing was the mammals living underground were protected and were able to survive and flourish over many millions of years. The mammals became the dominant species, and it's great they did, for at last check I believe we belong to that species also.

Another unique feature of earth worth mentioning is plate tectonic and continental drift. Caused by molten magna floating the earth's crust in many different directions from a point located in deep ocean trenches to a zone of subduction, where the continents return to the mantle to go through the process of melting and being forced up over and over again, this is the process called plate tectonics and is an ongoing and relatively slow process but is responsible for all surface features of earth, like mountains, continents, valleys, rifts, and basins, and is responsible for many other surface features.

The earth has also experienced two major ancient ice ages—one about 500 million years ago, and another about 250 million years ago, and also seven lesser ice ages approximately 50,000 years apart, the last one being about 10,000 years ago. The cause is difficult to prove but most certainly was due to some effect that caused the Earth to undergo periods of extremely cold temperatures. One can only imagine what these were. Could they have been changes in the earth's orbit or tilt? Maybe something affected the output of the sun. To explain these ice ages would require much more data than exists now. Since then the earth has been fairly stable with respect to temperature. One exception, though, is the recent melting of the polar caps in both the north and south. This, along with the melting of the glaciers, has caused great concern and has been referred to as global warming, the cause of which could be the cumulative effects of positions of the planets, their orbits, the moon, its inclination, and the presence of a tremendous increase of C02 (carbon dioxide) in the upper atmosphere, causing a greenhouse effect. These things are then responsible for the increase in the average temperature of our planet. The planetary effects we have no control over, which I believe is the primary cause and I estimate is 80% responsible, while the other 20%, I believe, is caused by man burning fossil fuels, such as forests, coal-fired boiler, petroleum-burning vehicles, which include cars, trucks, ships, jet planes, or anything using oil-based products. Consider also vapor trails of jets, of which there are 5000 flying in our atmosphere at any given second, and let's not forget the disruption of the layers of atmosphere each time we send a rocket to space—not only to the space stations, but also each time we send a research vehicle to outer space. If we care at all about our earth, we must find ways to stop this madness, or the inevitable result will be disastrous. I only hope we are not too late.

As of late, efforts are being made to produce photovoltaic panels to generate power. This is the way to go for not only scientific reasons, but also to provide jobs to help our failing economy. We must return to a manufacturing base for economic stability to restore our country to its former greatness, a point of concern. At the end of WW II, the U.S. was number one in the world and had the highest standard of living; we are now number eighteen with respect to the rest of the world. What happened? Could it be manufacturing sent offshore? What stupidity caused us to make this decision? I leave this for another book.

CHAPTER 5 – GALAXIES—DISTANCES AND TYPES

Galaxies—What are galaxies? They are concentrations of billions of stars grouped together in space. Each galaxy is separated from the others by thousands of light years. In order to understand what they are, let's examine our own Milky Way, which is a galaxy that is closer and can provide much more accurate data, so let's go! Our Milky Way galaxy is approximately 100,000 light years across, has a spiral shape, and a bulge at the center, which measures approximately 10,000 light years. Our sun, which is a member, lies 27,000 light years from galactic center, near one of the arms, containing many stellar objects of different ages, some of which are very old clusters and some new. The center of our galaxy is composed of fairly young stars—designated population I—and the outer arms are composed of older clusters and nebulae—designated population II. The density of the galactic center is such that it is quiet impossible to see through it, but we have been able to locate the approximate center. It was found to be near the constellations Sagittarius A.

Because of the rotation of our galaxy, it takes 225×10^6 years to make one revolution. It has been difficult to make this estimate because of differences detected with data gathered, which affect the expected orbital patterns. Some theories suggest that a controversial substance—dark matter—is causing gravitational differences in the data, but this imaginary substance is very controversial and does not have a big following amongst the scientific community. Later, in Chapter 8, I will address this issue in greater detail.

Our Miky Way galaxy can be easily observed with the naked eye by simply looking directly overhead on a clear night, away from light pollution. If you have never seen the Milky Way, it would be quite an experience, for it is a wonder to behold.

The nearest galaxy to ours is the large Magellanic Cloud, located 179×10^3 light years away and can be seen with the naked eye (if you're fortunate to live in the southern hemisphere). Two spiral galaxies, M31 (Andromeda) and M33 (Triangulum), both belong to the local group and are at 2.2×10^6 light years and 2.9×10^6 light years away. Beyond the local group, there are millions of more distant galaxies located in all directions. Their shapes, distances, and even their spectral compositions have been recorded and cataloged by many astronomers, one of the first being Charles Messier (1730- 1817). He cataloged 103 galaxies and in his honor, we use the Designation "M" as a prefix to many galaxies; some examples are M31 and M33, mentioned before. The Earl of Rose (1800 – 1867), using his newly constructed Giant Reflection telescope, which had a mirror six feet in diameter, was able to resolve for the first time M51's spiral shape, located in Canis Venatici. His original telescope is being reconstructed in memoriam and was to be working by late 2006.

Edwin Hubble, another great astronomer, made many contributions to science, which helped to better understand galaxies. One such contribution was his method of diagramming and classing galaxy types, which are used today. Their categories are: spirals, class Sa - Sc, barred spirals (-SBa - SBc), elliptical (Eo - Et), and irregulars, examples of which are M87 in Virgo, A type Eo, Spirals -NGC 7217 in Pegasus, Type SA-M81 in Ursa Major, type Sb-M33 in Triangulum Type Sc; also the barred spirals NGC 3504 in Leo Minor Type SBa, NGC 7449 in Pegasus Type SB6, and a SBc type in Hercules. The irregulars are not as well defined and tend to exhibit shapes of giant clouds or clusters. Besides having unique shapes, galaxies exhibit varying amounts of Red-Shift in their spectrum, depending on their distance. The deep space galaxy shows the greatest Red-Shift at between 1-15 billion light years because the Red-Shift is real, the cause has yet to be determined, and because recession doesn't hold promise with me, primarily because of the mystical state of dark matter that has not yet been proven to exist. I think the Red Shift has another cause (see chapter 8).

The galactic clouds such as Magellan are in themselves unique because they contain almost every type of celestial object—single giants, dwarf stars, double stars, novae, open and closed globular clusters, and gaseous and planetary nebula.

Cepheid galaxies such as M31 in Andromeda are especially important because they bring an additional tool—they help determine the distance, in light years, of distant galaxies to explain in the first place. They are considered variable galaxies; some of their components are stars of a variety of types), exhibit changes in brightness (intensity) with respect to time, and their rate of change in brightness has been well documented over many years. Their make-up (by spectra analysis) and their apparent brightness are used as measuring devices to improve the calculations to predict their distance. This methodology is one of the major inputs necessary in most cosmological formulas, which render extremely accurate measurement distances for each type of Cepheid galaxy.

Lastly, novae and supernovae are also used to predict galactic distances by comparing their parameters to those of similar local events and assuming their dynamic properties are approximately equal (luminous intensity - mass intensity and time of event) and then using these events, along with previously acquired data, gained from Cepheids and the Hertzsprung-Russell diagram to allow us to then make very accurate measurements of galactic distances.

In the next chapter, Chapter 6, I will endeavor to explain the history, discoveries, and how the Hertzsprung–Russell Diagram is used to predict the life cycle of all stars and stellar bodies. Basically, new stars are born on the upper left of the Hertzsprung–Russell Diagram and die on the lower right. So let's go to find how this useful tool (H/R diagram) was developed and how it is used.

CHAPTER 6 - HERTZSPRUNG-RUSSELL DIAGRAM

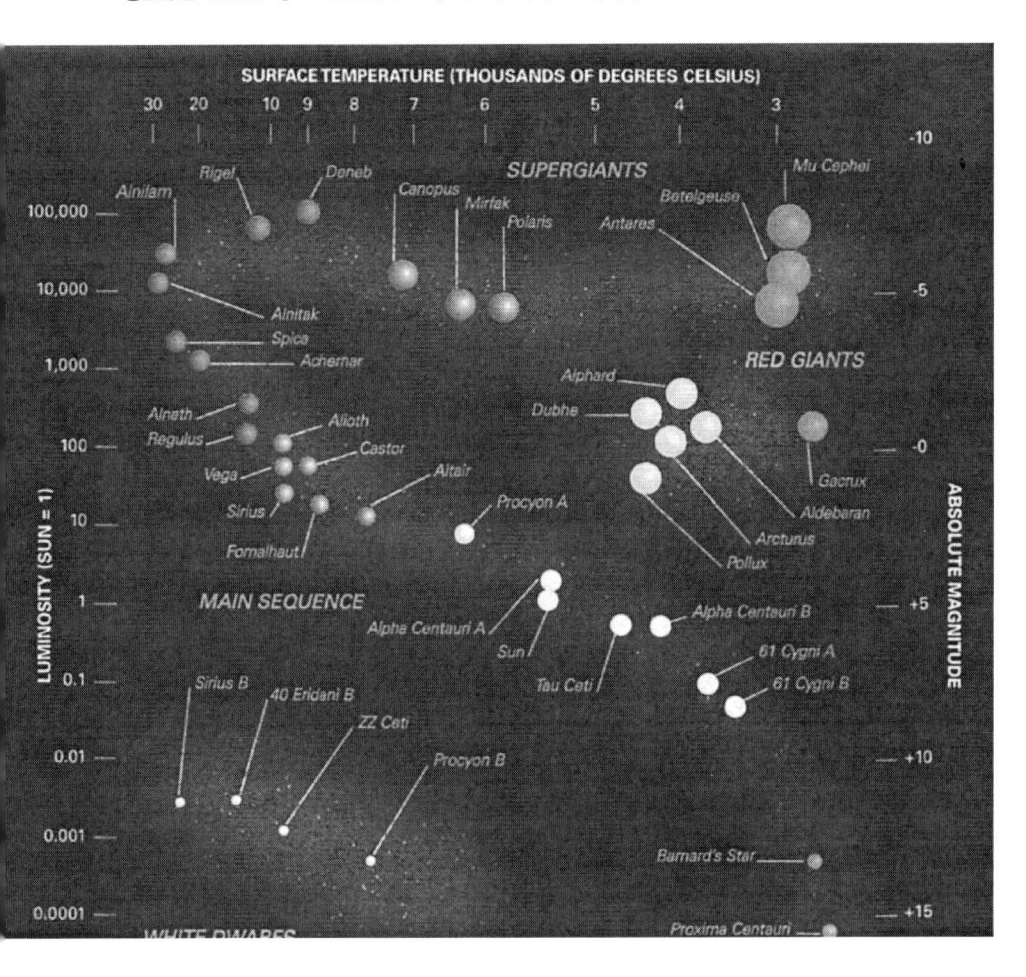

Two of the most important tools used in the field of astronomy are the telescope and the spectroscope. I will now give the reader my interpretation of how these instruments were used to develop the Hertzsprung-Russell Diagram. As previously mentioned, the telescope and spectroscope allowed astronomers to examine the components of stars at great distances. The spectroscope is an integral part of the telescope and is useless without it. I will describe the original spectroscope configuration, primarily because it is easier to understand. There are newer configurations of spectroscopes; although their principle of operation remains the same, they are far more complex in operations. Therefore I elect to describe the simpler configurations. The telescope focuses the incoming star light at a point called the Gould focus. The special construction requires prisms and slits to first break up then display the stars' fingerprints into a rainbow of colors from the low-frequency Red End to the high-frequency blue end of the light spectrum. These frequencies are measured in nanometers (nm), which is a very small part of a meter (1 x 10^{-9}). For example, the red light of the spectrum is approximately 700 nm, and this figure was expanded by Andres Jonas Angstrom, while doing his research in 1814 – 1874, by a factor of 10. Today we still use this reference because it improves the accuracy of the calculations; therefore, 700 nm = 7000 angstroms, or 7000 A because of the light of the distant source on its way as it passes through. Its upper layer contains the basic elements of which it is made, absorbing the light at that molecular vibration frequency. It appears as a dark line on the spectrum of that star. These dark lines are therefore referred to as absorption lines and thus reveal the composition of the elemental makeup of the star under study. After many years of using this method to catalog star types and categorize them accordingly, two astronomers from Harvard University in 1908, Ejnar Hertzsprung and Henry N. Russell, made an astounding discovery—the data they were compiling, when plotted on a curve, revealed a band from left to right that all stars fell into, and from this came the undeniable fact that all stars follow this curve. From this they realized they had discovered the life cycle of all stars—one of the most profound discoveries of modern astronomy. As the number of stars in their data grew, it became apparent that some means of keeping track of the star types was needed, so they came up with a method of naming the groups. In order from brightest blue (hottest) to less bright reds (coolest), this grouping was then given the designations W, O, B, A, F, G, K, M, Q, R, N, S. A way was needed to remember this sequence so a mnemonic was associated with it, which goes, Wow, oh be a fine girl, kiss me quick right now, sweetie.

    Each major star type was subdivided from 0 - 10 to increase the accuracy of the groups. For example, earth's spectral type is G2 and tells us where we belong, with respect to all other stars. The importance of the diagram is evident when we see just about every astronomer today making use of it in their calculations, or in papers written on the subject.

    One example of how the diagram can be used is when comparing two stars for identification as to their age and makeup.

Star number one is spectroscopically identified as belonging to type W on the upper left, and star number two is spectroscopically identified as belonging to groups S on the lower right. From this, their age, temperature size, and makeup are instantly derived. I'm sure the reader can appreciate how powerful a tool this diagram has become.

Another way the diagram is useful is, when plotting many stars, their population appears as a broad curve. As previously described, this curve is referred to as the main sequence. This information is then used to predict 99% of all stars, which accurately judges their normal life cycles, allowing us to estimate things like how old the earth is and how long our sun has before going to the stage of Red Giant. Another point became apparent when abnormal stars such as Super Giants and White Dwarfs are plotted on the diagram. Their place is immediately identified as an extreme case of birth and death, and their life can then be accurately predicted as being short lived.

Using this information, along with new telescopes both on earth and in space, to study deep space stars and galaxies, quasars and novae have given us a new insight never before realized, allowing us to modify existing data, help us develop new theories, and better understand the principles governing our universe.

Lastly, because of how convenient it is to use the H/R diagram, when a star's temperature is located on the diagram, this information can instantly be used with great confidence to predict the age of that stellar body. For example, when a star's age is found to be 10 billion years from birth to death, applying this fact to the so-called Big Bang age of the universe reveals an anomaly previously not talked about. How can an individual star be as old as the whole universe? Doesn't this present a lopsided scenario? Of course, if we consider the Big Bang in error, then this presents a whole different set of circumstances. I use this fact (see Chapter 8 and Chapter 9 references) to substantiate the problems that recession and the Big Bang Theory create.

I believe the reader would do well to consider the data, questions, and anomaly presented here and in Chapters 8 and 9. Many of these thoughts have come to my attention because of information acquired by use of analysis of spectral absorption lines and the data of this H/R diagram.

CHAPTER 7 – THE BEGINNING OF OUR UNIVERSE

The history of the idea of the beginning of our universe didn't gain popularity till after the advent of the Renaissance, somewhere between the fifteenth and sixteenth centuries A.D. The earliest theories were not very descriptive, were very general in nature, and required that the authors' concepts be taken at his word, without proof. There was a need to have some form of methodology in their writings, primarily because of the volumes of ideas and interpretations presented. Therefore, the Scientific Method was developed and is still used today. Here is a brief outline of how the Scientific Method is used to present or define any theoretical argument.
Observation – State the subject of your argument.
Formulation of your hypothesis – Explain your reasoning.
Experimentation to demonstrate your hypothesis – Show work.
Conclusion – Be sure it validates or modifies the theory.
    A. Sequency of generating your theory:
      1, Explanation thereof (tentative)
      2, Taken to be true for the sake of argument
      3. Proposals or suppositions
      4. Show any formulas of data to support your hypothesis.
    * This is the method I will use to support my theory in Chapter 8, The Red-Shift_(of distant galaxies).

There exist today two popular theories of how our universe began. The first is the Steady State Theory, presented by Fred Hoyle in 1932, and the second is the Big Bang Theory, presented by A. Hubble in 1936. Other theories have been proposed but by and large are contractions of the two above. Let's see which of the two make the most sense, keeping in mind the rules of the Scientific Method previously described. The Steady State Theory reasons that the universe is infinite and unbounded and that matter is neither created nor destroyed, but is simply in a state of constant change. (This to me makes good sense because it doesn't violate the laws of conservation of energy and matter). One example comes to mind— as a star reaches the end of its life and has used all its nuclear material, it explodes as many novae of sufficient mass do. Its material is then distributed into space, first condensing into gaseous nebulae and finally condensing further till its density is such that nuclear burning starts once again, and the star takes its place on the Hertzsprung-Russell Diagram as a new highly energetic young Blue

Star, to start to cycle all over again. By the way, I would like to mention that the process I have just described is the only process that can account for the heavy metals (such as iron, nickel, and gold, without which we could not exist) permeating our universe and cannot be accounted for by proponents of the Big Bang. Also, our Hubble telescope has recorded thousands of images describing the process precisely. I have just described this evidence I to justify my belief in the Steady State Theory, as proposed by Fred Hoyle in 1932. On the other hand, those who believe the Big Bang Theory as correct believe the universe is finite and bounded, and all matter came together some 10-15 billion years ago. At some undefined point in space it condensed to an infinitely small volume and then exploded with great force, going through various phases of expansion and ending up as the heavens, as we see them today. I have a number of problems with this theory. First, let's try to imaging this so-called Big Bang, especially the part about all known matter condensing at a single point in space, referred to as (To) (the time of the so-called explosion). My first question is where did the matter come from at time (T) = To (-) 1, before the event? I have heard many arguments defending the Big Bang all my life, not one of which makes any sense, and all disobey all known laws of science, as well as those of logic. One such theory, called the String Theory, referred to a state of nonexisting particles, strings combining to create matter. This argument makes no sense whatever, not only because they present no proof or hypothesis of the existence of these mystical strings, but their efforts to present a mathematical argument fail miserably because all scientists know that a mathematical argument without a practical example is useless, especially those having many mystical dimensions lacking any proof except their word. I challenge any of them to present a practical example of a universe based on five dimensions, needless to say six, seven, or more. Remember, mathematicians can say anything an author requires them to say, but without an example or proof, it is absolutely useless. I also get the feeling many of the Big Bang proponents can see problems with their initial ideas but have aligned themselves with it for so long and published so many books on the subject that they cannot bring themselves to admit they were wrong. This is the reason they continually come up with ridiculous proposals of states of nonexisting matter, such as violation number one—dark matter and dark energy, to explain their perception of what they think they see, such as recession of deep space objects (galaxies, quasars, novae, etc.) concerning the Red-Shift in absorption lines of those objects and the so-called forces pulling on them (see Chapter 8, where I go into the reason for the Red – Shift). Violation number two: The state all matter condensed to a single point in space at "To" must be the center of this event, yet when confronted with the question of where in space this point is, they give a nebulous answer that I don't think they themselves believe. They say the point is everywhere! Now, I don't know if they think they are being funny or if they believe they can say anything and convince anyone of anything for whatever egotistical reason, but remember, the burden of proof is on the one presenting the theory or argument, and it's up to them to prove their statements. Not only does the "everywhere" answer violate my sense of reason, it also defies logic because there is no way a single point can be everywhere at the same precise moment in time. Violate number three: According to the Big Bang guys, they all admit 90% of the matter's necessary mass must be the sort of matter we cannot see or measure. Then they mention dark mater and dark energy (Epicycles, anyone?)

Violation number four is at the limit of what they refer to as the envelope of the Big Bang, which, according to them, is somewhere between 10-15 billion light years distant and beyond which, nothing can exist! Well, how do you explain that Hubble's recent deep space images and the infrared Spitzer Telescope, launched in 2003, have detected galaxies extremely red-shifted, where none should exist (see chapter 8, "The Red-Shift"). Violation number five—the example of the expanding balloon, with particles within to represent the expanding universe. In this scenario they refuse to admit to a starting point of expansion, which is strange because, by their own admission, when they describe their idea of the beginning of the so-called Big Bang, they make reference to "To" at a point in space. I believe this is a deliberate exclusion because they see where the argument is going, and if they were to admit to a point in space, the next question is where in space is this point, followed by, if everything is receding away from us, does that mean we are at the center of the universe? Try to envision a point in an enclosed volume where all things are equidistant. This point can only be at the center. Now don't you feel uncomfortable thinking our earth is at the center of the universe. This convoluted explanation can only be the result of serious excuses contrived to defend their position. And for these reasons I find myself favoring Hoyle's Steady State Theory over the so-called Big Bang. Even Carl Sages, a proponent of Big

Bang, had reservations as to it being correct. In his book *Cosmos*, he questioned whether Big Bang was correct because of the missing matter (mass), and he also stated that there could be another reason not yet uncovered that could explain the Red-Shift (see Chapter 8).

Lastly, I don't believe the universe was created from a Big Bang, but I do think that continuing smaller bangs are possible, like the novae and supernovae or their remnants, quasars in deep space. This, as Fred Hoyle stated, is a continuing process with the consumption and creation of matter continuing into infinity. We must not try to overstep our acquired knowledge, but use it in such a manner that we have a good understanding of the subjects of our endeavors before we proceed to the next step. It is of prime importance to lay the groundwork, especially when introducing a new theory so there are no mystical reasons given to explain processes or convoluted arguments to substantiate a claim. Let's not forget Occam's razor: The simplest explanation is usually correct.

CHAPTER 8 - THE RED-SHIFT

In Chapter 6 I described how the spectrometer worked. Let's now consider what the shifting of the absorption lines of specific elements in distant galaxies means. It has been correctly detected, when observing the spectrums of galaxies at great distances, that the compositions of spectral elements are displaced to the red end (low frequency) of the spectrum when comparing those elements of the same type here on earth in the laboratory. For example, if we look at sodium lines in our lab, they appear at 589 nm, but these same lines in distant galaxies such a~s Virgo are Red-Shift, and the amount of red shift is dependent on how far away Virgo is approximately (50,000,000 light years). At this point I would like to mention there are two types of Red-Shift. The first type is rotational red shift. A component of distant galaxies, such as stars that exhibit comparatively small amounts of red shift because of the stars' rotation around the center of its galaxy. The second type of Red-Shift is much larger and causes a much greater displacement to the red end of the spectrum. This type of Red-Shift I will refer to as longitudinal Red-Shift and is a function of distance, not rotation. Although these two types of Red-Shift appear similar, the amount of red shift determines what category it belongs to. Edwin Hubble, as well as W. Shapely Ames, Humanson, and other astronomers in the early 1900's, were investigating rotation red-shifting of distant galaxies when Vesto Slipher discovered sodium lines displaced by great red-shifting not possible if caused by rotation. Hubble continued the study and also detected major shifting in absorption lines of many distant galaxies, which he interpreted as the source moving away from earth. In his mind he compared this observation to the Doppler Effect, which caused sound moving air on earth to change pitch (frequency). He was so sure of his findings, he visited Albert Einstein and told him of what he had found. Einstein, at first, was not impressed, but because of Hubble's constant insistence that he was correct and because he had convinced others in the astronomical community, along with the fact that at the time Einstein was deep in thought trying to put together his Unified Field Theory and had no time to devote to this unexpected diversion, he then allowed himself to be convinced to the point of dropping his universal constant out of his equation and stated it was probably the biggest blunder of his career. I personally don't think so. I still believe

Einstein's universal constant has merit, but I will leave this for a future discussion.

The question of what's causing the Red-Shift has been the source of many articles and much debate ever since its concept. At present the commonly accepted belief is the Doppler Effect caused by receding galaxies. But is not believed by all, including myself? This is the reason that motivated me to write this book. I will present my argument later in this chapter, but first I would like to reflect on how Hubble's idea became popular and also why I disagree with him. In order for Hubble's theory to work, certain assumptions had to be made. First his data showed the galaxies moving away not at a constant rate, but at an ever increasing rate.

His assumptions, no matter which direction you looked, created a paradox. His assumption was that there was some unknown force located beyond the detectable universe pulling at the galaxies, and this force increased in intensity with distance. This assumption is not only contrary logic, but also violates the principles of the conservation of energy and momentum. In order to explain this non sequitor, some astronomers as of late have introduced a mystical unproven force referred to as dark energy or dark matter, although they provide no evidence of its existence. This explanation does not sit very well with many in the astronomical community, including myself. Even Hubble, who proposed the theory that the Doppler Effect was responsible for the Red-Shift, questioned whether it was correct when applied to distant galaxies.

Recently an article called, "What's wrong with Recession Theory?" (*Astronomy* magazine), addressed a question worth considerating, questioning the speed of recession. For example, let's take the Sombrero Galaxy (M 104); its speed of recession is given as 2.4 million mph (= 700 miles/s), and this is a comparatively close galaxy, about 50×10^6 light years. The speed of recession increases when considering more distant galaxies. The question, then, is where the energy comes from to cause such acceleration. The present astronomical community all agrees, including myself. That matter, regardless of type—whether it be proton, neutron, or any other form at all, cannot produce acceleration; it must come from a force that's applied to the particles in question. No force, no acceleration, and not only acceleration but tremendous acceleration on the order of thousands of miles per second. Just to give the reader a feeling for this acceleration, consider that while launching our space shuttle, the ship approaches seven miles per second, not thousands. Once again this is in total violation of the conservation of energy and momentum, and with this I challenge them to present proof thereof. Could it be possible they are mistaken, and the galaxies in question are not receding at all? But the Red-Shift we observe has a completely different cause. In another article from the *Astronomical Journal*, the author presents his reasons why he discounts the idea that the universe is expanding. He reasons—and I agree—that if it were, then the density of space would decrease and the mutual gravitational attractions of particles would also decrease, then taking this argument to the limits the universe would eventually reach equilibrium or start to collapse. This is one of the anomalies I mentioned in Chapter 7 on the creation of our universe. Also, another point to consider refers to the density of space used by many astronomers. This has been my belief all along, and I cringe each time I hear someone refer to the vacuum of space. The reason why I have gone to great lengths to convince the reader of the density of space is because I use this fact in presenting my theory along with data. Let me state unequivocally that space is not a vacuum. It does have density, although the measurable amount is very small. This is not only reflected in my theory on Red–Shift, but it might also have been a contributing factor as to why the Michelson-Morley experiment failed. Their sample was far too small to be measured. I would also like to bring to the reader's attention that sometimes authors state ideas as fact in order to convince us of their way of thinking, especially with reference to dark energy or dark matter. I will remind the reader that they give us no proof, no definitions, no theories, and not even a hypothesis to justify their beliefs. None of this would be necessary if they admit to other explanations for the Red-Shift of distant galaxies.

Once again let me remind the reader that during the time of Ptolemy, "Epicycles" were invented to explain the retrograde motion of the planets. Everyone today agrees that "Epicycles" does not exist. Many papers have been presented over the years by many astronomers and scientists openly admitting there could very well be another explanation for the observed Red-Shift in the absorption lines of distant galaxies. But no one to date has come forth with an acceptable argument.

At this time I would like to present another explanation for the Red-Shift we observe. Suppose, instead of galaxies moving away at ever-increasing speed, their light wave's fronts are being delayed by the density of space. It's strange and also somewhat confusing, while reading Einstein's works on the Unified Field Theory, where in one chapter he makes mention of the vacuum of space (i.e. a vacuo), and in a later chapter he refers to the density of space having a positive number. Space cannot be both; it either has density or not. As I have stated previously, I say space does have density, although this density is very small per unit volume. When considering billions of light years, the cumulative effect becomes significant.

The propagation's delay that's caused by the density of space has a number of components, but the major component responsible for the delay is the name I have taken upon myself to coin. I call it the DiPhotonic Permittivity (Pr) of space because, to my knowledge, there has never been a mention to define or designate this effect. How I came to this conclusion was the similarity I saw of the parameters of photons moving through space and electrons moving through high-speed transmission lines when I was a research engineer and member of technical staff at GTE laboratories, and later for Siemens transmissions systems in Northlake, IL. Using this insight gained, there is what leads me to my ideas in applying the parameters of Giga Hz data to my astronomical background. The controlling parameters in transmission line was the Dielectric Permittivity (Er) and was responsible for the delay in transmission lines, so you can see how I connected the similarity between high-frequency propagations of electromagnetic waves and light waves in space.

It is now my intention to present my theory, using the Scientific Method previously mentioned, the method all astronomers and scientists are familiar with:

Observation:
>A detectable Red-Shift in the spectral absorption lines of specific elements in distant galaxies and other stellar bodies. This Red-Shift is presently thought of as being caused by the recession of the galaxies under study. I propose a completely different reason.

Formulation of your Hypothesis:
>I believe the Red-Shift in the spectral lines of distant galaxies are caused by impedance of the light wave fronts, which in turn are responsible for the propagation delay, and thus the Red-Shift, the primary component of which I refer to as the Diphotonic Permittivity and of which I assign the designation (Pr).

Experimentations to Demonstrate the Hypothesis:

I have chosen to use published data of five well known galaxies to support my theory and develop a methodology to predict and describe this propagation delay, incurred by the light wave front's understudy, and thus its Red-Shift. The data presented in Table 32 lists five galaxies, their distances from earth in light years, listed in nm and Å (Angstroms), as it appears in their spectrums.

Table 32

Galaxy	Distance in light years x10^6	Red Shift (nm)	Red Shift (Å)	Lab	Sodium Lines Observed (nm)	Lines Observed (Å)
Virgo	50	2	20	589	591	5910
Ursa Major	650	26.0	260	589	615	6150
Corona Borealis	940	37.6	376	589	626.6	6266
Boothes	1700	68.0	680	589	657.0	6570
Hydra	2700	108.0	1080	589	697.0	6970

Now, if Virgo's Sombrero Galaxy (M104) shows a Red-Shift of 20 A at 50×10^6 light years, it holds that for every 2.5×10^6 light years we will see a Red-Shift of one A. This delay is caused by the Dipotomic Permittivity (Pr) of space, which is the basis for my theory and is responsible for the observed Red-Shift. I have calculated a value of 0.4×10^{-6} Å per light year, associated with the delay, and have applied this number to the five galaxies in Table 32 to predict the Red-Shift of their sodium absorption lines in their Spectrum. I present this data as proof of my theory of how the Dipotonic Permittivity of Space causes the propagation delay that we see in the $\Delta\lambda$ of the galaxies' understudy and, to my knowledge, have never heretofore been mentioned. The $\Delta\lambda$ of each galaxy is easily calculated using my method and does not require convoluted ideas such as dark matter, dark energy, missing mass, or any other mystical unproven forces. They all disappear like the "Epicycles" of old, leaving us with simply the propagation delay of "c," the predicted values of the Red-Shift of the five galaxies, and others that I include in the Chapter 8 reference data pages. All the data presented was accurately proven to be true, within a range of ± 2%. Minor variations are possible because of the density of space; it is not homogeneous and does not have an absolute value per volume, especially when we consider such great distances of 50 million or 13.7 billion light years or more, but if a greater accuracy is thought to be necessary to prove the point, then a higher order of magnification can be applied to the data, which would yield a more precise result. But I think that the basic theory of propagation's delay caused by the density of space (Dipotonic Permittivity) is my primary motivation, not recession, by virtue of the Doppler Effect.

CHAPTER 9 – REFERENCE

To help the reader better understand Chapter 8, I have chosen to include reference material directly related to Chapter 8. The reference material was randomly selected and is in no particular order, but I'm sure it will help to clarify certain subjects and definitions. With this I present:

Space: Space is not a vacuum, does posses density, and has two measurable components—Capacitance © and Inductance (L). Although very small per unit volume on the order of $C = 10^{-12}$ parts / millions / light year and $L = 10^{-18}$/parts per million / light year, it is represented by $C = 1/4$ and $L = (p/6^2)$ and resonance $W^2 = \dfrac{4\pi Ne^z k}{m}$

These are just some of the basic formulas needed to do mathematical analysis which, for the sake of brevity, I must leave for a future study. But the reality exists that space does have density and a positive value, as stated by Einstein and which I have presented in my theory as having a value that is responsible for the Red-Shift of 0.4×10^{-6} A per light year and can be used to estimate the Red-Shift of other distant galaxies and stellar objects. Present-day astronomers and cosmologists use formulas such as $\dfrac{\Delta\lambda}{\lambda}$ to measure the amount of change detected in the Red-Shift of distant galaxies, an example of which is the Sambrero Galaxies (M104) in the constellation Virgo

$Z = \dfrac{591 nm}{589 nm} \cong 1.0034$

$-1 = 0.0034$,
and $\Delta\lambda = 2nm = 20\text{Å}$

at 50×10^{6} light years. The Red-Shift is an excellent metric to use to determine the distance of a stellar object. By the way, distance has always been a problem to predict and has changed values over the years. For many of the stellar objects under study for this reason, it's good to consider all available data, including Cepheid's, H/R Diagrams, Standard Candle data, and any other related data when predicting distance and Red-Shift. The more data used in your calculations, the better and more accurate the result.

It takes 500 sec or 8.33 minutes for an event on our sun to reach us on earth, like a violent surface explosion (sunspot). Even though our sun's distance (93 x 10^6 miles) is considerable, it's insignificant compared to the distance of, let's say, (M104), the sombrero galaxy in Virgo, which is = 50 x 10^6 light years away, and its light would take 50 million years to reach earth, assuming light could move through space at 186,293 miles/sec, unimpeded. But we know that space is not a vacuum and has density; therefore, space will restrict light in its travels. The fact that we see the Red-Shift of sodium lines in (M104)'s spectrum as it travels 50 x 10^6 light years to reach earth is proof of the propagation delay incurred, which is caused by the Diphotonic Permittivity properties I have discussed in my theory of Red-Shift.

Conversion Factors
To convert nm to Å, multiply by 10
Ex: Sodium lines 589 nm x 10 = 5890 Å

Some Convenient formulas to use
To find velocity
$V = \frac{D}{T}$ Where V = Velocity
              D = Distance
              T = Time

To find time
$T = \frac{D}{V}$ Where V = Velocity
              D = Distance
              T = Time

To find distance
$D = V * T$ Where V = Velocity
              D = Distance
              T = Time

Or time to frequency (Mega Hz – 10^6 Hz) or visa-versa, use the formulas f = 1/T and T = 1/F to change frames of reference, to simplify a calculation, or to better understand how one unit affects the others.

Ex: $f = \frac{1}{T}$ = 598 nm (sodium absorption line)

1.6977×10^6 or 1.7 Mega HZ and
$T = \frac{1}{f} = 1.7 \times 10^{\wedge}6$ Mega HZ = 589 nm or 5890Å

Diphotonic Permittivity (Pr)

    The component of space density that's responsible for the propagation delay of light, and which in itself has two subcomponents—a primary or static component, which is the dominant component, and a second or dynamic component, which can be thought of as having electromagnetic properties such as Inductive Reactance and Capacitive Reactance, the sum of which is responsible for the propagation delay that the density of space has on light waves moving through it. A value for the Diphotonic Permitivitly (Pr) has been determined to be (0.4 x 10^-6 Å /LY). Applying this value to a galaxy such as Sombrero (M104) yields a Red-Shift in Å of (0.4 x 10^-6) x (50 x 10^6) - 20 Å and applying this same formula to XMMJ2235, galaxy cluster at 9 billion light years yields a Red-Shift of (0.4 x 10^-6) x (9x10^9) - 3600 Å, which would cause the Red-Shift of sodium lines to be (5890 Å) + (3600 Å) = 9490 Å into the far infrared end of the spectrum and well beyond the visual limits of 7700 Å. What we must do when choosing a reference element absorption line is choose an element that has laboratory reference absorption lines far to the violet end, such that when shifted could easily be detected in the visual part of the spectrum. For this I have chosen ionized carbon (Carbon IV), which has a laboratory signature at 1657 Å Note (emission lines as well as absorption lines can be considered to be Red-Shifted in their spectrum by a similar amount.) Therefore, (1657 Å) + (3600 Å) = 5257 Å, well within the visual spectrum, and if we follow the above suggestions we would not be required to use complicated instruments such as infrared detectors and elaborate launch vehicles to be able to measure absorption or emission lines that have been shifted out of visual.

Below I will provide a list of the latest Hubble Ultra Deep Field Stellar objects as a reference to test my theory on Red-Shift.

HUDF Stellar Objects	R/S	Distance x 10^6 L.Y.7	Objects R/S Converted to % / C	R/S Converted to (A)	Total R/S Emission lines of Carbon IV
366	*	*	*	*	*
2457	*	*	*	*	*
1344	0.0369	504	0.369	2016	3673 (Å)
8316	0.62	5800	6.2	2320	3977 (Å)
7556	0.89	7600	8.9	3040	4697 (Å)
521	1.0	8000	10.0	3200	4857 (Å)
423	1.0	7809	10.0	3123.6	4780 (Å)
1446	2.47	11,100	24.7	4440	6097 (Å)
2881	4.6	12,400	46.0	4960	6617 (Å)
5225	5.2	12,600	52.0	5040	6697 (Å)
2225	5.8	12,760	58.0	5104	6761 (Å)
30591	6.7	13,000	67.0	5200	6857 (Å)

How was the speed of recession determined?

A reduction in the velocity of light was reported in the data collected by Slipher, Hubble, Humanson, Myall, and others in the absorption lines of sodium in the spectrum of distant galaxies, as stated previously. In order to explain this observation, they chose to interpret the Red-Shift as recession and proceeded to develop formulas to support their ideas. The most popular of which is $Z = \frac{\Delta \lambda}{\lambda}$. When they applied the formula to (M104) at (50 x 10^6 LY) distant, it yielded a figure of = (0.00403), and this figure expressed as a percentage of the speed of light equaled a value of 750.760 mps, as the following table shows.

To start, the speed of light "c" equals 186,293 mps
So 185,293 mps - 10% = 18,629 mps
186,293 mps - 1% = 1862.93 mps
186,293 mps - 0.5% = 931.465 mps
186,293 - 0.403% = 750.760 mps (Bingo!)

This is where the proponents get their recession figures.

I agree that the Red-Shift we observe of the sodium lines in (M104) spectrum at (50 x 10^6 LY) is shifted from 589 nm to 591 nm, but not by receding away from us (recession)—simply by attenuation of the light waves we receive caused by the density of space, which I have talked about previously and is the primary component of my theory (see pg. 71). I believe my theory is far more reasonable and have determined the attenuations (propagation delay) to have a value of (0.4 x 10^-6 Å/L.Y) x (50 X 10^6 L.Y.) = 20 Å.

How many miles does light travel in one year at 186,293mps?

186,293 mps x 60 sec = 11,177580 miles in 1 min
11,177580 mpm x 60 min = 670,654,800 miles in 1 hour
670,654,800 mph x 24 hrs = 16,095,715,200 miles in 1 day
16,095,715,200 mpd x 365.3 days = 5.88 x 10^12 miles in 1 year

So the distance light travels in 1 year is = 5.88 x 10^12 miles or 5.88 trillion miles or 6 x 10^12 miles in 1 year. Question: How many miles must light from (M104) Sombrero galaxy travel to reach earth? (M104's) distance is (50 x 10^6) x (6 x 10^12) = 300 x 10^18 miles. That 300 is followed by 18 zeros, or 300,000,000,000,000,000,000 miles. This gives the reader a feeling for how distant (M104) really is.

While reading Joe Silk's book *The Big Bang*, I discovered this

anomaly:
Comparison of Red-Shift Data

Galaxy	Life Book of the Universe 1972		Joe Silk's Big Bang Book 2001	
	Distance x 10^6 L.Y.	Recessions mps	Distance x 10^6 L.Y.	Recessions mps
Virgo	50	750	48.47	745.7
Ursa Major	650	9300	621.4	9321
Corono Borelis	940	13,400	870	13670
Booths	1700	24,400	1553	24,234
Hydra	2700	38,000	2461	37,905

Note the difference in the above data; on the left is Life book and on the right is Joe Silk's book. The source of the data was the same; the spectrographs were identical. The strange part is that the recession speed is very close, but the distance is not—why? There should be no difference. Is there a fudge factor involved? The original source of the data must be found; if not, then individual data of the five galaxies' understudy must be found.

We hear many times that the universe was very small many billions of years ago. This argument is based on the precondition of expansion and recession. The only way the universe could have been both very small and very large is in itself an anomaly based on the circular argument presented by Big Bang proponents (and by the way has always bothered me, for a number of reasons which I will endeavor to state now). The purpose of turning the clock backwards is to try to imagine what the universe looked like and its dimensions at time "To," the so-called beginning of the universe. Recently I read a book on Big Bang, which stated that at time "To," all matter in the universe was condensed to the size smaller then the head of a pin. This is absolutely not possible. Because all matter in the universe would have exploded long before coming anywhere close to the size of the head of a pin, and not even thousands of miles still would be insignificant, this gross misunderstanding renders their reverse projection meaningless. Lastly, the WMAP space probe has found the visible universe to consist of 4% matter and 96% empty space. This 96% has mistakenly been interpreted as missing matter because this is what the Big Bang boys need to justify their argument and refer to the empty space as 23% dark matter and 73% dark energy. How are you able to get a percentage of something that does not exist? This is just another example of a convoluted argument based on a total misunderstanding of the universe.

Photon mass (Mn)—The photon the fundamental particles normally thought of as the carrier of light waves and assumed to have zero mass (which, by the way, I have always disagreed with because zero equals nothing, and nothing is nonexistence) was recently found to have a photon mass of 7×10^{-17} ev (1×10^{-49} g) by experiments at Roderic Lakes at the University of Iowa. Using a modified Cavendish balance consisting of a steel toroid wrapped in current carrying coils, by observing the torque on the toroid duets, its dipole moment and its interaction with reference "$M\lambda^2 = /A/,$" which depends on the product (symbols), and estimating the vector potential, the photon mass was obtained. Assuming the photon mass is nonzero, this implies the speed of light is wavelength dependent and is affected by the electric and magnetic fields of space with no oscillations perpendicular to the direction of travel.

What was the diameter of the universe at (To)?

Is it possible to estimate the physical size of the universe at time equals zero (To)? In order to get a feel for the size of the universe at its beginning, according to the Big Bang boys, we must first run the clock backwards, then we start with the present estimate of 13.7 to 15 billion years. The one-way optical limit we can at present resolve. (By the way, this dimension represents the radius of a sphere, not its diameter, and if 13.7 - 15 billion is its radius, then two times the radius is equal to its diameter, or 27.4 - 30 billion light years!) I would like to hear an explanation by the Big Bang boys on that, but for now I will continue to estimate the size of the shrinking universe. We must now find how many seconds there are in one year. So in one year there are 60 s x 60 mins x 24 hrs x 365.3days = 31,561,920 s in 1 year (x) number of light years = 30×10^9. Diameter equals 9.468576×10^{17} seconds. Now to find the diameter of the universe at (To). We must divide the present diameter of the universe (30 billion light years) by the number of seconds in 1 year. This gives us the size (diameter) of the universe at 1 s. Diameter = $30 \times 10^9 / 9.4 \times 10^{17} = 3.19 \times 10^{-8}$ light years, and we must then find how many miles there are in 1 light year; there are 6×10^{12} miles in 1 light year. We must now multiply the condensed size, 3.19×10^{-8} LY's, by the number of miles, so $(3.19 \times 10^{-8}) \times (6 \times 10^{12}) = 1.914 \times 10^5$ miles, or 191,400 miles at T = 1 s, but we must continue to reduce the physical size because we are not at (To) yet. Therefore, I show a table of the reduction of the diameter of the universe.

A table of reduction of diameter of the universe with respect to time:

Time in sec	Diameter in miles 1 miles = 5280 ft
1	191,400
0.1	19,140
0.01	1914
0.001 (msec)	191.4
0.0001	19.14
0.00001	1.914
0.0000001 (msec)	0.1914 = 1010.592 ft

Now (5280 ft) x (0.1914) = 1010.592 ft in diameter, and at 1 usec after (To), the diameter of the universe is still 1010.591 ft. We still are not there yet (To). But just think for a moment—1010.592 ft is not the size of a head of a pin (a claim that the Big Bang boys flaunt). And all this matter in the observable universe crammed into a sphere 1010.592 ft in diameter? Not possible. It would have exploded long before reaching this point, and therefore the argument of all matter in the universe coming together at a point (To) is a fallacy and renders the Big Bang Theory a total misconception.

Other galaxies and their distance to disprove recession

Although I disagree with recession theory, I will endeavor to present the Big Bang argument, along with data, and in the process expose the faults of recession and find the truth of what is really causing the Red-Shift. In order to get a feeling for recession speeds of distant galaxies, we must extrapolate their data to the limit (the most distant) for galaxies under study. First taking their published data for (M104), the Sombrero galaxy, which is 750 mps, and this is equal to 1200 km/s less than "c" (300,000 km/s), we have 300,000 - 1200 = 298,800 km/s, which is equal to 0.4% at (50 x 10^6 LYs) for M104. Now, applying this formula to a table of other galaxies, we get:

Galaxy	Distance x 10^6 L.Ys	(formula)	Recession km/s	Velocity Difference %	R/S (nm)	R/S (Å)
M104	50	0.0034	1200	0.4	591	5910
SDSS galaxy	1,000	0.1	24,000	8.0	636.12	6361.2
XMMJ2235	9,000	1.4	216,000	72	1113.08	10,130.8
Qusar SDSS	12,000	5.8	288,000	96	1154.44	11,544.4

It makes absolutely no sense to assume a galaxy as massive as a Quasar SDSS, at 12 billion light years' distance, can be receding away from us at 96% the speed of light, and what was their reference element? It couldn't possibly have been sodium because it could have been shifted completely out of visibility (see Table 1 above), and, as stated previously, where does the force come from to accelerate such a mass to 96% that of "c"? They say they don't know; maybe it could be dark matter or dark energy, or maybe gremlins or leprechauns. "Epicycles" anyone? (Not good science!)

Now to get back to reality, consider the Red-Shift as a multiple of the Red-Shift of the reference galaxies M104 at (50 x 10^6 LYs).

There's no question that Quasar SDSS, at 12 billion light years, would show a greater Red-Shift of absorption lines in its spectrum. But we need to use a different element, not sodium, for the reason I give on page (it being shifted out of visual), but once we have chosen the element, the Red-Shift could be easily explained by my theory of attenuation per light year by virtue of the density of space and would not require any force at all, as the recession boys need. I present this as evidence as an argument to disprove recession.

Distant galaxies and their spectra

It has been noted, while examining spectra of certain distant galaxies, that a change in the Red-Shift of their absorption lines over a period of –one to two weeks has been recorded, is in no way related to their rotational Red-Shift data, and might be explained by saying they could be very dense dwarfs close to our solar system, but because there has been no significant change in their position in the sky over hundreds of years, and if recession were responsible for the extreme Red-Shift, it would be hard to explain how a galaxy's rate of recession could be changing by such orders of magnitude, first slowing to 750 mps then increasing to 27,000 mps in one week. This explanation makes no sense. Again, what force or power could possibly be responsible for accelerating such a great mass to such a great velocity in such a short period of time? This is probably one of the most convincing arguments against recession. But given the same scenario, and taking into account my theory of attenuation (propagation delay) by the density of space (Diphtonic Permitivity), it's far more likely that the density and gravitational fields the photonic wave fronts were passing through had variable parameters (not a constant value), which I discussed on pageThis could explain the difference.

Another thought on light speed

Many theories have been established on accelerating subatomic particles in cyclotrons, betatrons, and particle accelerators, both here in the U.S. and overseas, supposedly to increase our knowledge of what matter is made of and in the process, throughout the years, certain laws and rules were developed, one of which is the classical argument of when accelerating particles to near that of "c" (186,293 mps or 300,000 km/s), we could never reach "c" or speeds greater thereof because, and their argument goes, as the particle gets closer to "c," its mass increases, and theoretically at "c," its mass would become infinite and would therefore require infinite power, which is not possible, and therefore we could never accelerate any mass to "c" or beyond. But I wonder if they ever considered the fact that the particles being accelerated have no energy of their own; the only energy they have is being imparted to them by extremely dense electromagnetic fields requiring magnetic laws of coupling, and due to the fact that there is no perfect magnetic coupling (no losses), they could never accelerate a particle fast enough to exceed "c." What is needed to prove that an object can exceed "c" is to provide the test object with its own source of energy, such as a multistage space probe and accordingly, as its speed increases, its mass would decrease because of its fuel being used up. I then wonder how our present laws would stand up. I leave this to the future, sadly, because at present science plays second fiddle to politics. But wouldn't it be nice to spend 9 trillion dollars on R + D instead of wasting those dollars on supporting yachts, mansions, Bentley, and exotic lifestyles, or trying to educate people of different countries who have no desire to be brought up to our level of intellect? Remember the prime directive —noninterference—you cannot educate people against their will.

Lastly, a final thought on particles' mass exceeding "c." It has been noted recently that deep space, novae exhibits, and expansion parameters reflect the only explanation possible: The components in their outer parameter show evidence of speeds greater than that of "c." How could this be possible if no material object could ever exceed "c"? As they state!

What is $Z = \frac{\Delta \lambda}{\lambda}$ is the difference of an absorption spectral line of a given element, such as sodium. In the lab on earth, sodium absorption lines appear at 589 nm. Using the above formula, we can find the percent difference of the sodium line of a distant galaxy with respect to the sodium lines on earth. For example, Sombrero Galaxy (M104) in Virgo shows a Red-Shift to 591 nm for its sodium lines, and Z = 591nm/589 nm = 1.0034 (-) 1 = 0.0034, which is 0.34%; therefore, (M104's) light is slowed down by 2 mm = 0.34% = 1020 km/s or 298,980 k,/ and 300,000 km/s / 298,980 km/s = 1.0034. Therefore, Z represents the percentage difference of the Red-Shift of observed sodium absorption lines in the galaxy-wide study.

UNDERSTANDING THE UNIVERSE

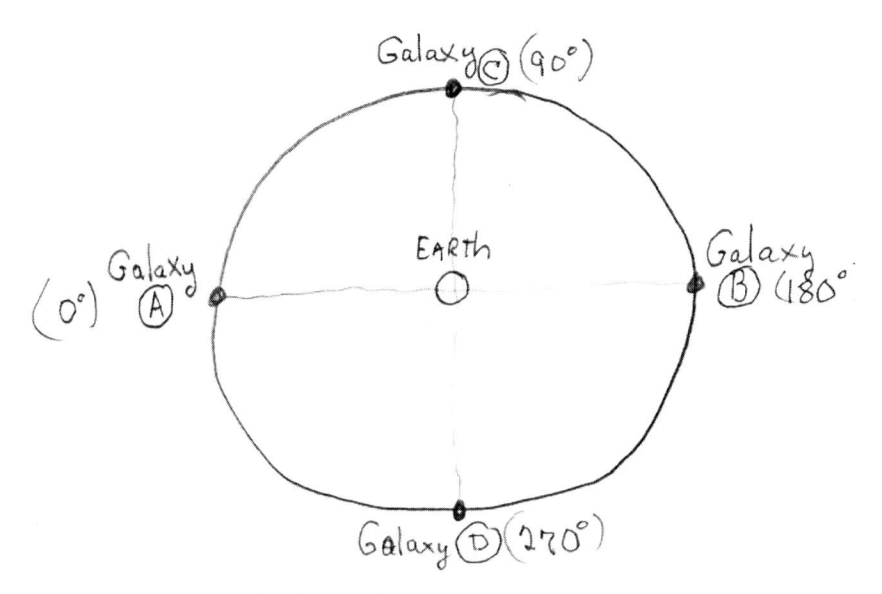

[Diagram: A circle with Earth at the center. Galaxy A (0°) on the left, Galaxy C (90°) at the top, Galaxy B (180°) on the right, Galaxy D (270°) at the bottom.]

Does Recession Make Sense?

Proponents of the Big Bang Theory say the universe is expanding (receding) away from us to a maximum distance of 15 billion light years in all directions, no matter which direction we look. Now let's consider if what they say makes any sense. They say if we see a galaxy A at 15 billion light years, and we rotate 180° to galaxy B, also at 15 billion light years, and we repeat this for galaxies C and D and measure 15 billion light years, and repeat this in every direction, no matter which direction we look, what does this scenario suggest? Answer: We must be at the center of the universe, or earth is the point at which our universe began. Can you possible believe this is true? Well, I don't! The scenario by itself should be all that's necessary to render the Big Bang incorrect. But there is more. Let's measure galaxy A at 0° and galaxy B at 180°. Their individual distances are 15 billion at 0° and 15 billion at 180°. Now measure point A to point B; this dimension is 30 billion. This contradicts the Big Bang guys and they know it; they proceed to explain this away with ridiculous convoluted arguments. They can't account for a universe mass of 15 billion light years, much less than one 30 billion light years; they openly admit there is something wrong. With their models they all agree that 90% of the matter (mass) of the universe is missing, and that's just the 15 billion light year model.

There's nothing missing, except possibly the missing gray matter between their ears—all their problems present themselves simply because they account for the $\Delta\lambda$ of distant galaxies, as receding away from us. And when asked what's causing this recession, they come up with the same old answer—a force called dark matter or dark energy. Unproven by their own admissions and mystical in nature, this is the same old circular argument getting us nowhere, and I'm surprised they don't see it. None of their contrivances would be necessary if they were to admit to another reason for Red - Shift (my theory at the beginning of Chapter 8). Once this is understood, all the other excuses could be thrown out and a simple distance/delay formula can be worked out for all stellar objects within our visual limits. This admission would also render Big Bang incorrect because no recession, no expansion, no Big Bang, and a universe that is infinite and unbounded.

Optical Visual Limits

We approach our visual limit somewhere between 10 - 15 billion light years. Light is shifted to the red end of the spectrum, to such a degree that we are no longer able to detect it in the visible part of the spectrum, and it appears to "wink out." (I discuss this in the next chapter), and if this is truly the case, then we can't see beyond that limit. In order for us to detect light beyond that limit, I can only think of two ways to prove the existence of stellar objects beyond, let's say, 15 billion light years. One is to move in the direction which we are observing, reducing the distance by some measurable factor—say 1 billion light years. This, I think, is quite impossible because we have not developed space travel yet to the point where we could send instruments to this distance to make the necessary measurements. Way number two would be to send an infrared space telescope out of our atmosphere, which would then be able to detect and measure accurately signatures of distant stellar objects in the infrared part of the spectrum, such as the Spitzer Space Probe. To document the existence of these objects would be a profound discovery, for it would prove beyond a shadow of a doubt that the Big Bang Theory was incorrect, and would lay to rest all the mystical excuses of recession and other contrivances we have heard for the last 50 years, such as String Theory, Super Sting Theory, or M Theory (the latest two-dimensional fantasy), which, by the way, were created to justify their position and not to aid in the understanding of true science. I believe it would do well to remember that mathematics, especially complex mathematics, can be used in such a manner to prove whatever the author would like, and can only be viable if a practical example can be shown.

Reflect back to the arguments of the flat earth theorist, before Columbus, or the argument presented by Ptolemy (Epicycles) to explain retrograde motion—how valid are they now!

The Crabe Nebula (M1) Super Novae

Observation of the crab nebula today shows an expansion of approximately six light years from the initial explosion on July 4th 1057 AD. It was recorded by the Chinese astrologer Dee Gue 953 years ago. This allows us to estimate the spread of expansion, which is the distance the remnant of the crab nebula has traveled since the explosion. It is given by the speed of expansion in miles per hour (2.5 x 10^6 mph), which yields (41,666 mps), which is 22.35% the speed of "c" at a distance of 6500 light years. The present size of (M1), the crab nebula is approximately equal to 6.0 light years, which gives 6.0 LYs/953 years = 0.063 LYs per year. So the force of the explosion was such that it caused the crab nebula to expand at a rate of 0.063 LYs per year. That is quiet an explosion.

Hubble's Recession Speed Anomaly Co-Moving Distance

I find it interesting that Slipher and Hubble, along with other astronomers of their time, came up with numbers for the difference in the Red-Shift of sodium lines for (M104), the Sombrero galaxy in Virgo. I explain on page 94 how I believe they came to this conclusion, but I also find it interesting when applying this same idea to other galaxies. No matter how far, the rate of change was always the same. Hubble chose to explain this Red-Shift by recession, but in recent years our ability to see far deeper into space and measure Red-Shift reveals an astounding fact that causes Hubble's original formulas to blow up, and the present astronomical community justifies this with a fudge factor called co-moving distance, the distance the galaxy under study has moved since its conception, which then yields a belief that these distance galaxies are accelerating away at an ever-increasing rate (i.e. the farther away, the faster they are receding). This dynamic action can't be explained by any of the proponents of recession, and when asked what could cause this, they say they don't know. That doesn't sound like good science to me, does it? I think if they consider my theory of attenuation, they might have a different idea of what is causing the Red-Shift. In my theory the cumulative effect of density (Pr) Diphtonics Permittivity is additive and therefore, the greater the distance the greater the Red-Shift, and it does not require any co-moving distance anomaly.

Propagation Delay of High-Speed Digital Pulses in a Transmission Line

In the laboratory, measurements are made to identify the parameters that control the dynamics of the medium (transmission lines) for a design of said systems. Ex:

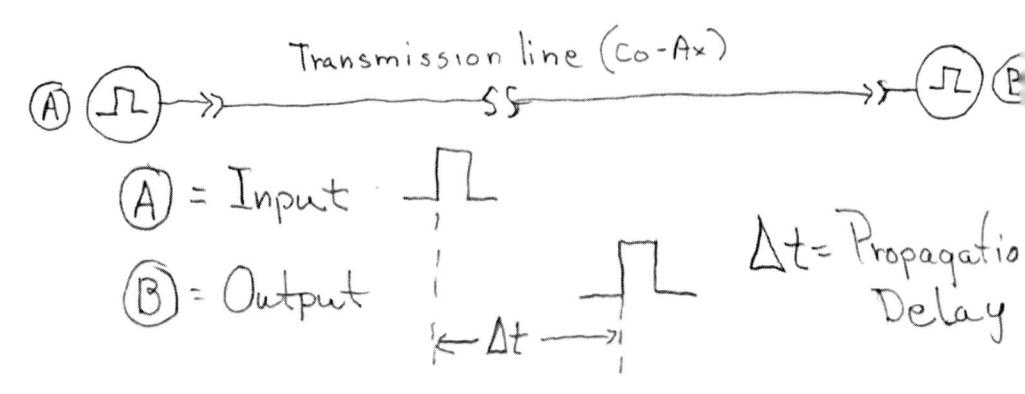

[Diagram: Transmission line (Co-Ax) from A (Input) to B (Output), showing pulse waveforms with Δt = Propagation Delay]

Ⓐ = Input
Ⓑ = Output
Δt = Propagation Delay

The velocity of propagation for all transmission lines is determined by the dielectric
permittivity, which is a function of the insulating material surrounding the center conductor.
Ex:

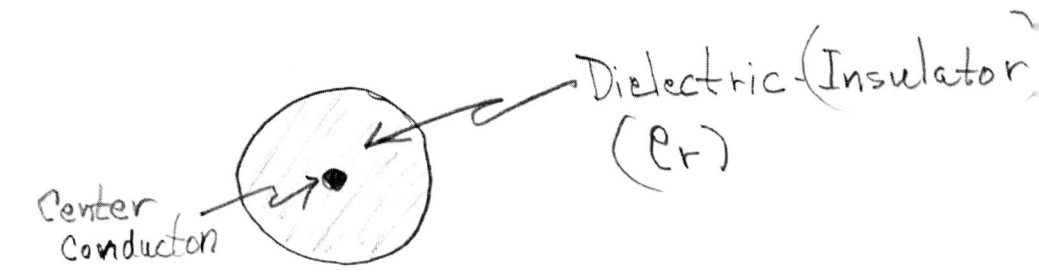

[Diagram: Cross-section showing Center Conductor and Dielectric (Insulator) (ε_r)]

Different insulating material have different (er), such as the following examples in Table
Dielectric materials are numerous; here are just a few examples:

UNDERSTANDING THE UNIVERSE

Material	er
Guta Purcha	1.5
Polystyrene	1.2
Polyester	1.1
Air (Vacuum)	1.0

The greater the density, the greater the (er), and the greater the effect the dielectric material has on the propagation delay. Electromagnetic waves have been shown to move at the speed of light in a perfect vacuum, as presented in the table above. Similarly, photons have been shown to exhibit the same properties as electrons when considering propagation velocity (which, by the way, Einstein received the Nobel Prize for in 1905 (the photoelectric effect), but space is not a vacuum and does posses density. As I have stated before, this density is what is responsible for the Red-Shift we see in the absorption lines of various elements and in the spectrums of distant galaxies as shown below.

Ex:

① (A) — Small distance → (B)
Sodium lines
f = 589.0 nM

Sodium lines
f = 589.0000001 nm

No change in velocity can be measured because the distance is far too small, but if we consider

Ex:

② (A) — Large distance → (B)
Sodium lines
f = 589.0 nm

Sodium lines
f = 591.053 nm

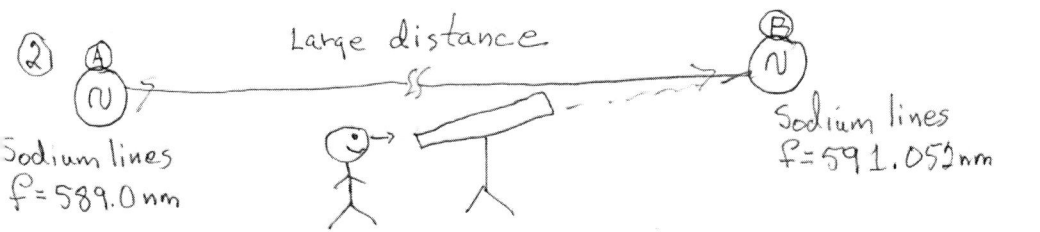

In example number two, a telescope and spectroscope are needed. A definite Red-Shift of sodium absorption lines is detected, and the amount of Red-Shift is a function of the distance of the galaxy under study, and therefore it follows that if the distance is (50×10^6 LY), or if the distance is (10×10^9 LYs), the ($\Delta\lambda$) of "c" is "C" - (Pr), where (Pr) is the Diphotonic Permittivity of space and has a value of (0.4×10^{-6} Å/LY), and taking the greater distance as an example we have (10×10^9 LYs) x (0.4×10^{-6} Å/LY), and this equals (4000Å) at 10 billion LYs. So you can expect to see a Red-Shift of the sodium absorption lines of a galaxy at that distance from (5890Å) + (4000Å) = 9890Å, well beyond visual (7000Å) and into the far infrared part of the spectrum. It would be convenient for us to choose a different reference element, such that after exhibiting the same Red-Shift = 4000Å, we would then have an absorption line that's in the visible part of the spectrum.

CMB The Cosmic Background Radiation

The cosmic background radiation has been talked about as being the remnants of the Big Bang and yet is measured beyond the envelope of the so-called Big Bang. I wonder if anyone has ever thought of CMB as the signature of deep space electromagnetic noise, caused by the infinite number of galaxies and other stellar bodies that exist far beyond, let's say, 15 billion light years that are going novae in a continuous cycle of birth and death. A profound question needs to be asked: Is the known universe younger than the stars and galaxies it's composed of, or are these stellar objects older than the so-called Big Bang? If Big Bang were correct, then neither A nor B could be true. The only way A or B could be true is if Steady State is correct.

The Increasing Age of the Universe

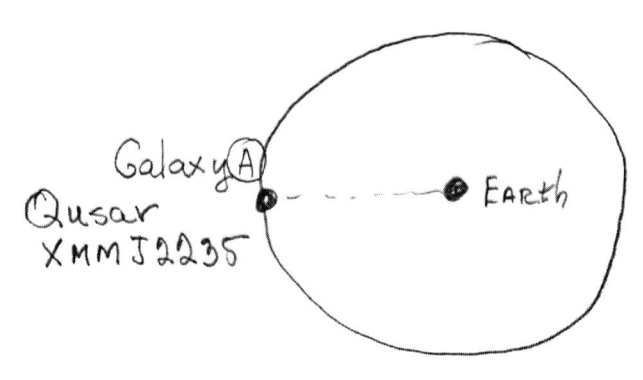

An additional argument comes to mind: If galaxy A, for instance, is receding away from us at 75% the speed of light, and that recession increases with distance, then it holds that at some time in the past, this recession speed was much less. For example, what was the theoretical speed at 10 billion or 5 billion or 1 billion years ago? Using their logic, this would mean that sometime in the past it took a longer time to cover the distance than it does today, thereby increasing the age of the universe by that amount. Now, because this would mean a universe 2-3 or 4 times older than 15 billion years, I think this in itself would render the concept of Big Bang totally impossible, the reason being that taking a universe 60 billion years or older (and once again, the Big Bang boys can't account for 90% of the matter of our present universe to be 13.7 billion years old). I would like to see how they twist the facts about this. A second thought comes to mind: If this so-called expansion were to continue in all directions out to infinity, then how is their universe any different than Hoyle's steady state? Their arguments are meaningless and restricted in thinking. Once again I submit my theory (propagation delay). No convoluted explanations are necessary to explain R/S. What you see is what you get.

In 1925 an astronomer wannabe by the name of Fritz Zwicky made mention of a tired light theory and suggested that light was slowed down by gravity—strange! But before I found my solution (Pr), I too referred to the Red-Shift as my tired light theory, and thought it was original, Zwicky's solution (gravity), although possible to have an effect on light propagation, I believe is far too small to account for the tremendous Red-Shift of distant stellar objects.

The age of the stars and galaxies in the universe are in disagreement with the so-called age of the Big Bang universe primarily because how can you possibly explain the fact that distant stellar objects exhibit parts of their spectrum, suggesting they are older than the universe at that time and distance, and this is only a problem because the Big Bang boys suggest that the universe had an absolute beginning (To), none of which would be true if we see the universe as in a constant state of birth and death, the conservation of mass and energy, or Hoyle's Steady State Theory. There would be no anomaly because stellar bodies, no matter how far distant, can be any age depending on when they were born! I believe it was Visto Slipher in 1912 who started this whole thing by mistakenly measuring nebulae which (unbeknownst to him) were actually galaxies, and then interpreting their Red-Shift as a Doppler Effect, not as caused by attenuation of the density of space, as I believe.

There are a number of opposite point of view between the Big Bang boys and myself—let's investigate.

Big Bang Boys	What I believe
Red-Shift – Recession	Attenuation
Origin of Galaxies – a false assumption that Qusars are young galaxies	Galactic distance reveals varying ages of stellar objects – what we see is what we get!
CMB – Mistaken belief its an echo of the Big Bang	Radiation emission from all matter from an infinite universe
Age of stellar bodies – convoluted contradiction that distant stellar objects are younger then the universe	Proof that the universe is in a constant state of evolution, beginning (birth), and ending (death)

And the final question, that of creation, is the ultimate question. Before we can begin to answer this one, we need to advance our knowledge considerably. Just think of asking a caveman (assuming you had a time machine and could speak his language) to explain gravity. For myself, I believe in a supreme creator, for the reason I have stated before in detail, namely, where did all matter come from at (To) = (-) 1 sec? Although this same question is also true of an infinite universe, we cannot answer it by any conventional scientific methods or theories put forth to date, simply because we have not advanced enough in our intellectual capabilities. If we continue to improve our knowledge in one or two thousand years, we will possibly be able to answer that question. And remember, infinity is a lot different than 13.7 billion years.

A Thought on Time Dilatation

Dating back to the 1800's, many physicists introduced a factor called time dilation, which suggested time measured was a function of distance traveled with respect to observers and because of this, coined terms like space time and frames of reference. Many of the early empirical examples that were presented to prove their beliefs have been shown recently, to be nothing more than a convoluted argument based on the notion that certain laws of physics were true. But using reverse logic has proven to be false, leaving the question of time dilation yet to be proven and the argument of clocks in orbit not absolutely devoid of preconditions. I leave this subject for future discussion.

There are many unanswered questions about Red-Shift of distant stellar sources, such as quasars. One which comes to mind if two different stellar objects are observed at the same distance and their Red-Shift was due to recession, then their absorption lines should be equal, Hal Arp, and other astronomers have found that there are some stellar objects that share common visual details and yet have completely different Red-Shifts, which suggests that this difference is not due to recession, which also suggests that the Red-Shift is caused by something else. I must throw my hat in the ring with other astronomers who have spent their lives contributing to the better understanding of how our universe really works. Some of them I'm sure you recognize: Geoffrey Burbidge, Sir Fred Hoyle, Hermann Bondi, and Thomas Gold, just to name a few.

Another point I would like to make is, because of the tremendous distances, there is no absolute Scientific Method we know of to accurately predict the distance of deep space stellar objects. Because of this, we estimate their distance by using parameters such as brightness, the H/R diagram, Cepheid's Variables, and spectroscopy. As I have stated before, the more data we have, the more accurate can be our conclusion, and we should not have to resort to any mystical, unproven contrivances such as dark matter and dark energy. Using data and the Scientific Method, in my mind, is far more professional than unproven and mystical ideas to solve problems and derive an understanding that is more accurate in conveying the true reason for a particular phenomenon.

<u>Distant Galaxies' Age Difference</u>

If our Solar System and earth are approximately 5 billion years old, and a distant galaxy such as SDSS Quasar is thought to be 13.7 billion years old, then:

13.7 Billion (SDSS)

<u>- 5.0 Billion</u>
8.7 Billion Difference from SDSS

Quasar was born 8.7 billion years before our Solar System and earth, and, according to the Big Bang boys, moved away from us by that distance 8.7 billion light years; therefore, it must have moved approximately 9 billion light years in 9 billion years. Thus, one light year per year is the definition of how far an object can move at the speed of light, and if one light year is equal to 5.88×10^{12} miles at "c" (186,293 mps), we have $(5.88 \times 10^{12}) \times (9 \times 10^{9}) = 5.292 \times 10^{22}$ miles moved.

Pf: $\dfrac{5.292 \times 10^{22}}{5.88 \times 12^{12}} = (9 \times 10^{9})$ or 9 billion light years

Therefore, in order for SDSS Quasar to show a Red-Shift of $Z = 5.8$ due to recession, it must have moved at the speed of light for 9 billion years. I don't think this is possible because what force could possibly be responsible for moving such a massive stellar body, the size of SDSS Quasar, over such a distance for so long a time? And for this reason I discount the idea that recession is responsible. As I have stated in my theory, the Red-Shift we observe is not a function of recession, but is caused by attenuation, i.e. the Diphotonic Permittivity (Pr) of the density of space and does not require a belief in any mystical, convoluted, unnproven, nonexistent forces such as dark matter, dark energy, strings, super strings, M-Theory, or any other multidimensional mathematical states of nonexistent matter. Let's revisit the definition of propagation delay and propagation velocity. Propagation delay is the time it takes for a signal to go a distance from point A to point B. Propagation velocity is the speed at which the signal propagates, and, for electrons and photons, is equal to the speed of light only in a vacuum. And because we know space is not a vacuum and does have density, the propagation velocity is reduced by this density.

Lastly, the term "look back time" can be simplified as nothing more than propagation delay (see above).

CHAPTER 10 - UNDERSTANDING GALAXIES

What if instead of a Big Bang, as discussed in Chapter 7, the universe is composed of an infinite number of local groups, which in turn are composed of billions of galaxies going through various stages of evolution, and this process continuing on into infinity? The latest images of deep space galaxies show tremendous age differences, some older than the so-called age of the Big Bang. This could not be possible unless the Steady State Theory were true.

I think galaxies have circular motion because there is a continuity of all matter in the universe. If we were to examine all elements existing in nature, starting at the subatomic level, then atomic, molecular, solar, and galactic level, a motion is imparted between particles because of their charge (+) or (-), and this charge sets up a state of directional motion (not random), and, because of gravitational attraction, causes this motion to become captured and circular. This moment of inertia is then transferred to the next level up, in order of complexity, until finally reaching its highest macro level, that of galaxies, maintaining the law of conservation of energy and momentum.

According to proponents of Big Bang, all galaxies were 100% completed very early in their history of the universe. But individual stars show ages of 10 - 15 billion years old, and others 25 million, many located within the same cluster. According to Big Bang, they should all be approximately the same age, and yet this fact cannot be accounted for. I have heard arguments referring to calendar age and generic age being responsible, but this to me is an admission that this difference in age does exist, regardless of what name is given. Instead of defending their beliefs, they should try to explain their logic with good scientific evidence, not rhetoric. These arguments make no sense at all; it's like listening to a Republican telling us how good the U.S. is doing while the stock market collapses. According to their logic, all galaxies should be 10 billion years old, or there should be no galaxies younger than 2 billion. This realization of age difference should in itself be reason for the Big Bang boys to think again about how sound their logic really is.

During analysis of Hubble's deep space images of galaxies, it was noted that at the extreme distance of 10 -15 billion light years, there could still be seen, beyond those resolved, other faint galaxies that were Red-Shifted to the point that their total output appeared in the red portion of the spectrum, between 6000 - 7500 Å, suggesting to me that those galaxies, so Red-Shifted, were on the verge of what I like to refer to as "winking out." If this is truly the case, then we are at the outer limits of visual observation, and even if there were other galaxies beyond, let's say, 20 billion light years, we would not be able to see them because they have Red-Shifted out of visual and into the far infrared. I wonder if this is what has prompted the surge in the number of infrared space probes in recent years, such as Spitzer. This idea, "winking out," is unique in nature and to my knowledge has never been mentioned before, and, in one respect, could be thought of as viewing the light barrier. A way to verify the extreme shifting would be to choose a reference element (discussed in Chapter 8), that, after experiencing extreme Red-Shifting, could still be detected in the visual part of the spectrum, and therefore would not require the use of expensive equipment and space probes to go out beyond our atmosphere (which causes attenuation of the infrared signal).

CHAPTER 11 - FUTURE OF PLANET EARTH

The future of Planet Earth can best be estimated by observing our history. Estimates put the beginning of our Solar System at 5 billion years ago, and earth at 4.7 billion years ago. It took another billion years to cool and acquire water; that puts us at about 3.7 billion years. The age of earth between then and now comes to us via geological and archeological discoveries. We know that the sun is a G2 type star and has a stable life of 10 billion years. According to Hertzsprung–Russell, that means our sun has another 5 billion years to exist before it expires, but that's if we don't destroy ourselves first by nuclear war, overpopulation, collision with some other stellar body, or let's not forget poisoning earth and ourselves with toxins or destroying the ecology, causing global warming by releasing excess amounts of CO_2 into the atmosphere. Surviving these disasters, our sun would determine the final outcome. The 5 billion years stated above for how many years of stable life our sun has left does not take into account its final stages of demise, long before our sun runs out of hydrogen fuel to burn. Its surface would first start to collapse; the next step would be when the core temperature reaches 40 million degrees and starts burning helium, at which time the sun would start to expand and become a Red Giant. This would continue till it encompassed the earth. The estimated time for this to happen is 3 billion years from now, but the oceans would have boiled off long before that—say, 1 billion years, so that leaves us about 2 billion years to develop our warp drive and get the heck out of here.

CHAPTER 12 - POSSIBILITY OF LIFE OUT THERE

Do I believe there is other intelligent life out there in space? The answer is a resounding No! Let me tell you why. In my 70+ years on Earth, I must have heard every story you could image, read thousands of books on the subject, including *Project Blue Book* from the U.S. Air force, watched hundreds of movies and TV specials, and attended many lectures, all of which presented no factual evidence of there ever being anything remotely considered to be intelligent life. Let's also include the Setti project at Arecibo in Puerto Rico which, by the way, lasted for 50 years and not one thing was ever heard.

In Chapter 2 I gave the reasons why I believe our earth is unique in the universe and is the only planet capable of supporting life, and I will ask those who think otherwise to present their evidence. The biggest argument in my favor is that not one single artifact or communication has ever been discovered or presented to substantiate the existence of life, not to mention that of intelligent life. Many authors write books and make movies and TV shows for one reason only—not to increase man's knowledge, but to make money! I have never seen a TV program about UFO's that ever showed anything of value to support their claims. The space ships were always out of focus, the characters were not real, the so-called evidence consisted of pie pans, double exposures computerized imaging, or a combination of all of the above. The so-called evidence is a joke! I will continue to keep an open mind on the subject. But remember, the burden of proof is on the person telling the story.

CHAPTER 13 - PROBLEM WITH SPACE TRAVEL

When I was a young man, I would daydream about humans landing on the moon. That was in the 1940's; my visions were about 20 years early. As time went by, we finally landed men on the moon, and many things became obvious. Space was very inhospitable to humans. It wasn't just the obvious things like the lack of air and water, but more subtle things like how gravity affects our bodies, such as bone loss, heart and circulation problems, and disastrous mechanical problems like generators exploding, meteors puncturing the hull of the space ships, or the danger of running out of air, or freezing to death, and let's not forget radiation. There is no atmosphere in space to protect us. And theses were the problems we incurred just going to the moon and the space station. The next step of setting up an outpost on the moon will certainly present a whole set of new risks, and going to Mars will increase that danger by a factor of 100 because of the time spent away from earth. Plus, there are the dangers of equipment breaking down and the inhospitable conditions on Mars itself. New technologies will need to be developed. First, a new form of propulsion needs to be invented, along with artificial gravity and inertial damping systems; methods to protect the space ship while navigating through asteroid fields, such as force fields and shields, would also be needed. Does this sound like *Star Trek*? Oh yes; while we're at it we might as well mention we could use warp drive also. I will remind the reader that none of the above have yet been invented, so where are we? Thinking about traveling to other star systems? Forget it. The nearest star is Proxima Centuri, 4.2 light years away. Once light years is 5.87×10^{12} miles, so 4.2 light years is $24,654 \times 10^{12}$ miles, that is, 24,654 with 12 zeros behind it. Now consider if we could travel at 10,000 miles an hour—it would take us 281,000,000 years just to get there. But what would we do then? Proxima Centri has no planets to land on, so what do you do now? Turn around and come home, so before you start, you must have a purpose and plan in mind, and if your propose is to land on a planet, the distance would have to increase by at least a factor of 100, and this in itself is no guarantee because we still do not have the capabilities to be able to see that far in space to resolve planets in orbit around other stars. So now I hope you see what asking for the future of space exploration involves. None of the things I discussed would be possible without the inventions mentioned. And now we come to the Coup deGrace: Who's supposed to pay the bill? Our government would have to change totally from politics as usual to a program of science and R&D for the benefit of mankind, and even this would be no guarantee that the necessary inventions would be developed. The reason is that you can't just throw money at an idea and think that would produce results, although it would be a step in the right direction I don't mean to discourage optimistic ideas about our future; I am only looking at the practical side. So if you are of the science-minded type, let's go to work. We have a lot to do.

GLOSSARY

A

Absorption line - Dark line in spectrum
Accretion - Debris field condensing
Apogee - Greatest distance from center of an elliptical orbit
Attenuation - Offer resistance to
Atom - Smallest unit of an element
Aurora - Sun's radiation colliding with earth's upper atmosphere

B

Background Radiation (CMB) - Radiation coming from deep space
Baryon - Basic components of atomic nuclei, protons, neutrons
Big Bang - An imaginary theory that requires the belief in unproven ideas such as recession, expansion, dark matter, dark energy, and (To)
Big Crunch - Universe collapsing on itself
Binary Star - Two star systems
Black Body - A body that emits all light absorbed on its surface
Black Dwarf - A star that has collapsed on itself after burning all its nuclear fuel—very dense
Black Hole - An imaginary place in the universe where all things fall into because of its extreme gravity
B.L. Lacerate Object - A variable stellar object at great distance which exhibits no absorption or emission lines
Blue Shift - A shift in the spectrum frequency due to the source moving toward the observer
Bode's Law - An empirical relationship between the distance of a planet from its sun, by J.E. Bobe (1771)

C

Cepheid - A variable star that shows rate of change of brightness and is used to estimate its distance
Celestial Equator - A projection of earth's equator on the heavens
Center of Mass - The absolute center of any body
CCD - Charged Coupled Device - a semiconductor whose electrical characteristic changes with respect to the light

falling on it

Chromosphere - The layer of the sun between the photosphere and the corona

Cosmological Constant – Einstein's universal constant, which he looked for all his life and has not been found yet!

Cosmology - The study of the universe

Cosmogony - The study of the evolution of the universe

D

Density - Mass with respect to volume

Doppler Effect - Change is frequency (pitch) of sound as it moves toward or away from a person

E

Electron - Subatomic component in orbit around its nucleus charge (-)

Ellipse - An oval curved path around two foci

Emission Line - A white line in the spectrum

Escape Velocity - The speed needed to escape the gravitational attraction of a body

F

Frequency - The number of cycles of a wave motion past a point in one second

Fusion - An atomic process in which the elemental nuclei are joined together, releasing a larger amount of energy in the process

G

Galaxy - A large grouping of stars, gas, clouds, and dust, containing million or trillions of components and ranging in size from thousands to hundreds of thousands of light years in diameter

Gravity waves - The theoretical force (attractive) that's necessary for two or more objects to pull together

H

<u>Heliocentric</u> - The sun at the center

<u>Helium Burning</u> - The atomic fusion that converts helium to carbon and oxygen, taking place after hydrogen burning and just before the star goes to a Red Giant

<u>Hertzsprung-Russell Diagram</u> - A plot showing the birth and death of a star's life cycle, with luminosity plotted vertically and temperature horizontally

<u>Hydrogen Burning</u> - An atomic fusion that converts hydrogen to helium, which powers our sun

I

<u>Infrared Radiation</u> - An electromagnetic wave having a frequency in the 6000 – 7000 Å part of the visible spectrum

K

<u>Kepler's Laws</u> - Describe planetary motion: 1.) All planets orbit in an ellipse. 2.) The orbital velocity is a variable. 3.) The time of an orbit is equal to its average distance.

<u>Kuiper Belt</u> - A zone existing beyond the orbit of Pluto where many comets gather

L

<u>Light Year</u> - The distance light travels in one year without attenuation, equal to approximately 5.88×10^{12} miles

<u>Local Group</u> - Nebulae or bodies associated with a galaxy at a distance, closer to the center

<u>Luminosity</u> - The light energy emitted in one second from a star or stellar body, expressed in watts (our sun emits 3.8×10^{26} watts, as an example)

M

<u>Magnetic Field</u> - An area surrounding a dipode magnetic body with reference to planets, usually suggests a liquid metal/molten core

<u>Magnification</u> - The increase in visual size created with the aid of a

telescope, described by the following formulas: M - f1/f2, where M = Magnification, f1 = focal length of the objective, and f2 = focal length of the ocular

Main Sequence - A curve plot on the Hertzsprung-Russell diagram, showing new young bright stars on the upper left and old dim stars on the lower right—the left are blue and the right are red

Messier Catalog - A catalog published in 1781 by Charles Messier, listing galaxies and clusters with designation from M1 to M103

Meteor - The designation given to a meteoroid when it enters earth's atmosphere

Meteorite - The designation given to a meteoroid when it reaches earth's surface

Milky Way – A galaxy of which our sun is a part and comprises vast members of other stellar bodies of different ages

Moon - A satellite such as our moon in orbit around its planet

N

Nebula - A cloud of dust and gas in space

Newton's Laws - Published in 1687: 1.) A body at rest remains at rest, and a body in motion remains in motion. 2.) Force applied to a body causes acceleration in that direction. 3.) For every action there is an equal but opposite reaction.

Nova - An exploding star, usually a white dwarf, noted by an extremely bright increase in light output of a thousand times or more

O

Ort Cloud - A band of icy planetesimals and comets ranging in size, located between 1 and 1.6 light years away from our sun

Open Cluster - A loose cluster of stars numbering up to a few thousand

Opposition - A planet's position when it's opposite earth, on the other side of the sun, and appears highest in the sky at midnight

Orbit - The circular path a planet takes around the sun when equilibrium between the planet's mass and velocity are exactly balanced with its mutual gravitational attractions

P

<u>Parsec</u> - The distance a star has in an annual parallax of 1 second of arc, equivalent to 3.26 light years or 19,200 billion miles
<u>Perigee</u> - A point in which a body in elliptical orbit is closest to its focus point
<u>Photon</u> - An individual packet of electromagnetic energy riding on a wave of light
<u>Planet</u> - A body in orbit around a star that has much less mass and no light of its own
<u>Planetesimal</u> - A considerably smaller body than the smallest planet in a system
<u>Plasma</u> - A state of matter that can only exist at extremely high temperatures, which causes the elemental components to be in an ionized gaseous state
<u>Precession</u> - A conical orbital pattern scribed by extension of the axis of a planet—for example, one that has a wobble or not a perfectly true spin
<u>Proton</u>- A part of every atomic nucleus having a positive (+) charge and consisting of 3 quarks
<u>Pulsar</u> - A rapidly rotating neutron star, emitting tremendous energy from both poles

Q

<u>Quark</u> - A fundamental particle that is the basic part of all atomic nuclei, consisting of groups of 3 to make up baryons and paired with anti-quarks to form mesons
<u>Quasar</u> - Very massive, extremely dense, and at great distance, on the order of billions of light years, and of which are also extremely powerful emitters of light and other spectral sources

R

<u>Radial Velocity</u> - A velocity that is measured around a central point
<u>Red Giant</u> - A star that has reached the end stage of hydrogen burning and starts to swell; it is located on the lower right of the Hertzsprung-Russell diagram
<u>Red-Shift</u> - The displacement of spectral lines of elements to the low-

frequency end of the visual spectrum

Reflecting Telescope-An indirectly viewed telescope whose image is reflected back from a concave mirror to an optical lens

Refracting Telescope - A directly viewed telescope whose image is passed through an objective lens and then through an ocular lens for magnification; it was the first type of telescope

Relativity - Einstein's theory that all events in nature are relative to their frame of reference

Resonance - A sustained regenerative oscillator

Retrograde Motion - The appearance in the sky of a change in the direction of motion of, let's say, a planet, actually caused by the position of observation with respect to the background stars and the difference in orbital position and speed, and also the source of one of the greatest misunderstood anomalies of science; Ptolemy referred to it as "Epicycles," which did not exist, the so-call secondary planetary orbits

S

Satellite - A moon of a planet

Seyfert Galaxies - A very bright galaxy that exhibits fluctuations and is used as a source to measure distance

Sidereal Time - The apparent time of the rotation of the observed heavens actually caused by the earth's rotation, equal to approximately 24 hours

Solar Flare - A violent release of energy on our sun, like a sunspot

Space Time - Proposed by Minkowksi in 1908 and became part of Einstien's theories to bring the reality to light of a fourth dimension

Spectral Class - A method of classifying stars according to the Hertzsprung–Russell diagram, having the designation O, B, A, F, G, K, M, K, R, N, and S

Spectroscope - A device used to display the components of the visual spectrum, which appears as a rainbow of colors when white light is passed through a prism; also used to identify a stellar body's component make-up by examining its fingerprints deployed as absorption lines of elements

Star - A massive stellar body which emits light energy by nuclear fusion

Sunspot - A dark spot on the sun's surface caused by the cooling, which is the aftermath of a solar flare

Supernova - An event of such magnitude that the explosion causes an increase in brightness of a million times, usually thought to be caused by the collapse of a White Dwarf

T

Tectonic Plates - Large sections of earth's crust of earth's mantle, moved by capillary actions created by molten magna flow, resulting in all surface features of earth

U

Ultraviolet - The part of the visual spectrum that exists at approximately 400 nm

V

Visual - The part of the spectrum from violet at 400 nm to red, at 700 nm

W

WIMP - Weakly Interactive Massive Particle - Theoretical elementary particles that are assumed to have a mass thousands of times that of a proton, but at present have not been proven to exist

White Dwarf - A collapsed star which has used all its nuclear material and is radiating energy purely by gravitational compression, and which is approaching the end of its life

Wolf-Rayet Star - An extremely hot star surrounded by gas, and shows strong emission lines in its spectrum

X

X-Rays - Electromagnetic frequencies that exist in the range of 0.1 to 100 Å—below the visual spectrum

Y

Year - The time it takes for a planet to complete one revolution around its star

Z

<u>Zenith</u> - A point directly overhead at 90°